Emmanuel Levinas on the Priority of Ethics

Emmanuel Levinas on the Priority of Ethics

Putting Ethics First

Joshua James Shaw

CAMBRIA
PRESS
Amherst, New York

Requests for permission should be directed to:
permissions@cambriapress.com, or mailed to:
Cambria Press
20 Northpointe Parkway, Suite 188
Amherst, NY 14228.

Library of Congress Cataloging-in-Publication Data

Shaw, Joshua James.
Emmanuel Levinas on the priority of ethics : putting ethics first / Joshua James Shaw.
p. cm.
Includes bibliographical references and index.
ISBN 978-1-60497-521-5 (alk. paper)
1. Levinas, Emmanuel. 2. Ethics. I. Title.

B2430.L484S525 2008
170.92—dc22

2008008317

TABLE OF CONTENTS

Foreword

As I sit down to write these words of introduction to Joshua Shaw's excellent book on Emmanuel Levinas, the morning's news is filled, as it regularly is, with stories about the plight of the victims of floods and tornadoes, about the hopes and suffering of war-torn peoples, about violence in neighborhoods and schools, and in general of the troubled, uncertain lives we live in our society and elsewhere around the world. The twentieth century saw the invention of modern warfare and a scale of violence and catastrophe previously unknown, and we are its heirs. For all their successes, modern science, advanced technology, and bureaucratic sophistication have provided us with nothing less than more efficient tools to injure and annihilate; and religion, for all its advocacy of decency and humanity, has also cultivated fanaticisms that live on the violent hatred of those deemed enemies of what is holy and right. It may be that the vocabulary of crisis is overused, but who can deny that it is appropriate. We are in a bad state and one that does not promise to abate.

In such a situation, it is tempting to believe that what is required to renew our lives—individually and collectively—is a kind of perception or knowledge that we have somehow lost, or, alternatively, that we need to recover emotions and feelings that we no longer have. Indeed, it may be that we require both, that our problems arise from both cognitive and motivational or attitudinal deficiencies, and that we therefore need to acquire a vision or understanding and a set of feelings or attitudes that have been overwhelmed by views and emotions that threaten to destroy us and our civilization. On such a diagnosis, our plight requires therapy; we need to be cured of our ignorance and our emotional defects, and what this means is education and affective training.

As I see it, this is the human and historical situation with which Emmanuel Levinas sought to cope. The aftermath of World War One, the rise of fascism, the horrors of Nazism and Stalinism, the incomparable threat of annihilation associated with Hiroshima, and all the atrocities and catastrophes that followed in their wake shaped Levinas's world and engaged him as a philosopher and as a human being. In philosophy—which probably was conveyed to him by Dostoyevsky and Tolstoy before he studied Bergson, Husserl, and Heidegger—Levinas found a path to a response to a world plagued by violence and suffering. But it was a conception of philosophy not as doctrine or method, rather it was a conception of philosophy as a way of living that arises from a critical engagement with alternative philosophical approaches, both those of the past and those of the present. It is not philosophy as science or philosophy as doctrine; it is philosophy as conduct, as living with others in the world. It is philosophy as ethics.

Western philosophy, one might argue, has always had links with religion on the one hand and science on the other. Viewed in one way, Levinas's work seeks to reestablish its connection with religion after a period, at least since logical positivism, when philosophy's primary association has been with science and especially the natural sciences. Viewed in another way, Levinas can be seen as someone for whom philosophy's most intimate association is with ethics, in the broad sense of all that is relevant to human character and social life, someone for whom ethics is more primary than either science or religion—or art, for that matter. However one understands Levinas's central teaching, it is not sufficient to state it in such elliptical ways. What is needed is a deep and probing examination of what he says in order to clarify its meaning, to show how Levinas reaches out to the ethical and the religious in very compelling ways, and that he does so in order to help us to reorient our lives and perhaps to find hope in the midst of destruction and despair.

Joshua Shaw's wonderful book on ethics and politics in Levinas's work seeks to do just this. Unlike so many of Levinas's interpreters, Shaw writes with stunning clarity and grace, and in an engaging and accessible way. Reading Shaw, one is not dismayed by Levinas's obscure terminology or confused by his views; rather one is drawn into a process of thinking about serious and deep problems. It is a methodical yet appealing itinerary of raising questions, proposing and considering solutions, revising and debating, that invites the reader to just the kind of attentiveness and responsiveness that Levinas takes to be fundamental to all of our experience and our lives. The reader cannot help but be enlightened and provoked by Shaw's patient and thoughtful prose. It is a combination of perspicuousness and philosophical depth that is rare in writing about Levinas.

Shaw raises a central question: does Levinas, for all his attention to what he calls the ethical, avoid concrete engagement with justice, with our ethical and political lives? Has he been correctly criticized as abstract and irrelevant? Indeed, is the very idea of his giving ethical advice and counsel and providing an account of justice somehow incoherent? Once he has clarified exactly what this challenge means, Shaw proceeds to defend Levinas against this charge and indeed, in the course of so doing, to defend a reading of Levinas that he calls "pragmatic." The argument he provides will no doubt find its opponents. As he points out, many believe that for Levinas the ethical is inexpressible and hence no articulation of it for practical purposes is possible. But Shaw's argument is subtle and nuanced, attentive to what Levinas says and also to the philosophical issues involved; it deserves careful scrutiny. As he sees it, Levinas does offer a justification for the ethical character of the particular relation each of us has to each and every other person; for Levinas the ethical arises first, so to speak, in our interpersonal relations, and from it advice about justice can and does flow, albeit in highly particularized contexts. If we generalize about what is required, such generalizations are rules of thumb or recommended policies. Deriving them and then acting with respect to them and the demands of the particular situation require deliberation and reflection and an appreciation of what is demanded of each of us. There is no recipe for how to act in a given type of situation, but there is much that one can and ought to do to determine what would be caring and just.

In the course of developing his argument and his view of Levinas, Shaw draws on a variety of interlocutors, some the continental figures who are considered members of his own philosophical tradition, others members of the analytic or Anglo-American tradition which he did not know. In this respect, Shaw's book is indeed special. Only a few books and articles attempt to place Levinas in conversation with figures such as John Rawls, Bernard Williams, Charles Taylor, and Hilary Putnam. Shaw does this in a natural and effective way, for these figures are introduced not simply to draw a comparison or to make a contrast; rather they are called upon to address objections to Levinas or to offer alternative answers to questions raised of him. These philosophical voices, then, become participants in the drama of the argument, so to speak, and the reader is thereby shown how he or she too can—and might profitably—enter into a similar dialogue with Levinas and his work.

In the end, Shaw underlines a point about Levinas that is powerful and important. Much of Western philosophy is grounded in ontology or epistemology. Its starting point is an account of what there is or what we know. Shaw's book argues that for Levinas, ethics is primary, but more importantly he shows what that means. He shows, that is, what it means to say that Goodness comes before Being and Truth. Levinas often points out that Plato in the *Republic* saw that this was so, but it is one thing for Plato to have said that the Good is beyond being and the ground of being, it is another for us to be able to understand this priority. As Shaw shows, for Levinas every relationship we have with every other person is grounded in the preciousness of that other person to us, and that preciousness requires our acknowledgment and support before it does or says anything else to us. *Ought* is prior to *is*, the good to the true. How we understand the world around us is determined by what matters to us, and what matters to us first and foremost are other people. Levinas, I think, wants us to appreciate this about our lives in part because it is what characterizes all social life, but also in part because we ought to live this way. Indeed, living in such a way that we are indeed responsive to others is the ground of any hopes we might have that the next century will be able to alter the tendencies that we have inherited from our recent past. Our best hope of overcoming atrocity and suffering, poverty and need, is to be summoned by the face of the other person and to respond to that summons with care and concern.

Professor Michael L. Morgan
Indiana University

Acknowledgments

I owe a great deal to many friends and colleagues. The person to whom I am most indebted is Michael Morgan. I began to study Levinas in Mike's first seminar on Levinas at Indiana University. I had tried to read Levinas several years, but it was only under Mike's guidance that I began to make progress in understanding him. For years, Mike and I met weekly to read and discuss him, and we continued to share our ideas via e-mail long after I was no longer a graduate student. Mike's book, *Discovering Levinas*, exemplifies what I hope to achieve in my own writing, both in its clarity, its precision, and the way it ignores the supposed boundary between analytic and continental philosophy. Mike has been a constant source of guidance, both as a mentor and as a friend, and I want to thank him for his support and for contributing the Foreword to this book.

Portions of this book began as a dissertation project, and I was blessed with an extremely supportive yet critical dissertation committee. Paul Spade mentored me for several years while I was a graduate student, and I credit him with inspiring, in no small measure, my philosophic sensibility. I have never met anyone who so deftly unites a passion for questions, a passion for paradoxes, and a passion for all that makes life, for lack of a better word, *fun* than Paul. I also want to thank Paul Eisenberg, who prepared copious notes and criticisms for, as I recall, every page of the manuscript that evolved into this book.

Portions of this book were presented at conferences, and I benefited from the feedback I received from participants. I greatly benefited from presenting material at a conference on Levinas and Wittgenstein at Indiana University. My thoughts on Levinas were enriched from conversations with Jeff Kosky, David Finkelstein, Simon Critchley, Alice Crary, Simon Glendenning, and Gary Gutting. Here I need to single out Simon Critchley for special thanks. There are several moments in this book when I criticize Critchley's writings on Levinas. The fact that I so often turn to him to formulate my own ideas indicates, however, just how pivotal a role he has played in the scholarship on Levinas. If I take issue with him, it is in terms I borrow from him. I am indebted to him for bringing to articulation so many of the debates I address in this book.

Finally, I want to thank Cambria Press. My sense is that the support and attention that Cambria gives their writers is truly exceptional in the world of academic publishing. Toni Tan went out of her way to keep me posted as to the progress of my manuscript and to recruit referees for it. I also want to thank the outstanding anonymous reader for his or her excellent responses. This book is vastly improved from the revisions he or she suggested.

Prologue

Letting Levinas Wake Us Up

To think is no longer to contemplate but to commit oneself, to be engulfed by that which one thinks, to be involved.

—Levinas, "Is Ontology Fundamental?"

1. Why This Book Worried Me

I found myself worrying, as I wrote this book, that its point became less clear the harder I tried to explain it. I imagine that most writers experience this worry, but I found it particularly worrisome.

First, one of my goals is to be especially clear. Emmanuel Levinas's writings are notoriously difficult, riddled with quasi-technical terms, rhetorical flourishes, and deliberately paradoxical turns of phrase. Many commentaries borrow his terminology without making much of an effort to paraphrase it, so that they are only marginally more accessible than Levinas himself, despite their claims to clarify him. I do not necessarily mind this obscurity, but I found it alienating when I first read Levinas. I recall feeling as if I had discovered someone whose work seemed vital yet impenetrable, and I remember seeking out literature on him with uncharacteristic desperation. I felt I had a duty to understand what he was saying about ethical responsibility, but could not do it on my own. I often found myself just as perplexed by the secondary literature on him and the familiarity with contemporary European philosophy it assumed. Since then, I have had much time to study Levinas and recent European philosophy, and I no longer feel this frustration. Yet I do worry that the esotericism of some of the scholarship on Levinas prevents his writings from achieving the influence they deserve, and one of my aims is to explain his ideas as clearly as I can. So, it troubled me that my book's point seemed to become more distant the harder I tried to clarify it.

However, I had a more significant reason for worrying. What drew me to Levinas was the note of moral urgency in his writings. I had the sense when I first read him that he was issuing a personal call to responsibility to me, his reader. A friend tipped me off to him as I was graduating from college, and I read *Totality and Infinity* for the first time while living as a graduate student in Chicago. I recall reading the discussion of morality and violence in the preface to *Totality and Infinity* one evening and feeling compelled to put the book down and to stare out the windows of my apartment. I could not have explained what I had read, but I felt I understood that it had said *something* about the need to put the book down, to stare out those windows, and to honestly face the world on the other side of them—a world where people are victimized far too routinely; where, for example, one out of every three women will be sexually assaulted; where one out of every three African American men will be imprisoned; where, chances are, the clothes you are wearing as you read this were sewn by a woman who did not receive enough money for her labor to buy herself the food she needs to survive; where each of us, it often seems, can excuse our indifference if we are eloquent enough to craft an elegant political justification or stubborn enough to pretend not to notice; and where the only thing able to block this process is the fragile awareness that each of us is incomparably precious and that some incalculable crime is done whenever any of us suffers.

The other philosophy books on my desk seemed shallow. Not intellectually shallow. Many of them were, I still believe, far more intellectually rigorous than *Totality and Infinity*. It was just that they seemed complacent in their failure to address themselves to the world outside those windows. It was hard to return to them without feeling guilt. It was not that I felt like I was selling out in reading them, that the only proper response would be to put aside philosophy as elitist or self-indulgent and to become a full-time activist. It was more that the stakes of doing philosophy had been clarified. The fact was that the world beyond those windows would be there if I chose to recognize it or not, and it changed the moral stakes of studying those books. It changed everything. To borrow from Levinas, it "de-neutralized" the safe, insulated world of my apartment.[1] I was still free to read my books and to write essays on them, but there was a glaring question as to the value of this writing. Would it be a tool for producing indifference? Was I using it to distract myself from the world beyond those windows or to excuse myself for not doing more to intervene in it? Would it allow others to evade their responsibility? Or would it attempt to intervene in that world, if only by sharpening my own sensitivities to the suffering of others?

The Levinas I admire is the Levinas I intuited on that first night of reading him. This Levinas is defined by his insistence on the absolute need to help those who are vulnerable. He is concerned with sensitizing his readers to their calls for help, and the chief philosophic puzzle that obsesses him is how we manage to ignore these calls. Why aren't we stopped in our tracks by statistics like those quoted above? What intellectual crutches do we rely on to convince ourselves that we are not responsible? In what ways have various philosophic systems been complicit in abetting such self-exculpation?

I think that many are drawn to this Levinas. My sense from talking with other Levinas scholars is that several had similar "conversion experiences" the first time they read him. I find, however, that this emphasis on the need to help others is oddly absent from the secondary literature, where there is a tendency, instead, to dwell on broadly epistemological debates over the limits of language, comprehension, and the consistency of Levinas's critiques of other philosophers. This shift in focus is somewhat understandable. Simon Critchley asks in the *Cambridge Companion to Levinas* whether "ethics [is] the right word to describe the experience that [Levinas] is trying to express?"[2] Critchley's contention that Levinas does not have a true theory of justice is a target of criticism in this book, but it would be obtuse to deny that he has a point. For all his talk about ethics, Levinas does *not* say much about what norms should regulate conduct; he *does* spend a lot of time discussing language and comprehension. It makes sense that much of the literature should follow him in focusing on epistemological issues as well.

Yet it is difficult not to feel that something important has been lost. Here I find it useful to borrow a distinction from a colleague who told me that I was interested in the "spirit" and not the "letter" of Levinas's philosophy. His own view was that Levinas makes the same antifoundationalist points about language and Western metaphysics as does Jacques Derrida, but that Derrida explores them with greater subtlety. He added that he knew a lot of people like me—people who admire Levinas because they perceive him as having something valuable to say about the need to help others. For his part, he did not see it. When he turned to Levinas's writings, once he got past the rhetoric about ethics, all he found was a simplistic form of deconstruction. I suppose that one way to describe this book would be as an attempt to do justice to the "spirit" of Levinas's philosophy by showing that it is present in its "letter" as well. Levinas does talk a lot about language, comprehension, and Western metaphysics, but the driving force behind his claims, I argue, is his conviction that we cannot help but acknowledge others as having an unconditional claim on our care and support.

I will have more to say about this thesis. For now, I just want to point out why my concerns about this book's clarity troubled me. I found myself worrying as I wrote it that I may have replicated the sin I was railing against. At the foundation, my goal is to share a reading of Levinas that does justice to his claims about the priority of ethics. The Levinas I admire is the one who tells us that nothing matters more in this world than helping others. My worry is that this emphasis on service to others has gotten lost in the secondary literature, where the tendency is to focus on epistemological issues instead. This book tries to correct this error by showing how Levinas's epistemological insights flow out of his claims about ethics, hence the title *Putting Ethics First.* However, there is a difference between establishing this *interpretive point* and the larger insight it is intended to convey. For me, Levinas's novelty consists in how thoroughly he appreciates how urgent it is to help those who are vulnerable. From this perspective, many of the debates over whether he violates his own claims about the limits of language and comprehension strike me as missing the point, which is precisely to set aside such theoretical debates in order to actually help those in need. I believe that I succeed in establishing this book's interpretive point, but I worry that in doing so I may have fallen into this pattern of focusing too heavily on the theoretical at the expense of the ethical.

So, what I wish to do in this prologue is allow myself to indulge in a different style of writing. I try to be clear throughout this book, but I do write in the voice of a philosopher who is concerned to defend an interpretation of Levinas. I think that I succeed in establishing this interpretation, but I worry that in the process, I may have failed to say enough about the book's larger point in an appropriately emotional or passionate tone of voice. For now, then, I want to take a step back and allow myself to write in a looser, more direct way about what I hope to show. To do this, I want to say something about what I think Levinas has in mind when he talks about "face-to-face encounters with the other." The face-to-face encounter with the other is arguably the cornerstone of his philosophy. Here, then, is the best example I can give of an encounter.

2. Face-to-Face Encounters on Interstate 84

It is December 24, 1999. My father is driving home along I-84 toward Brookfield, Connecticut. Here is how I picture him in my mind.

He is happy. Ordinarily, he loathes this part of the day; it is the worst part of an unpleasant job. He works as a tool and dye maker at a woodworking mill in Danbury, Connecticut. The hours are long. He works six days a week. Over the years, he has acquired a hacking cough from the sawdust and metal shavings he has swallowed. He is partially deaf from the roar of the machinery. His hands are riddled with scars. But, it is the long drive home in bumper-to-bumper traffic that he hates the most. In a few years, he will treat himself to a new truck, one with a CD player and air conditioning, and the commute will be bearable for a few more years. However, on the day of this story, he is still driving a ramshackle Nissan pickup: no air conditioning, no radio. The truck is barely functional; its frame is all but rusted away.

But, he is happy. First, it is the end of the workweek and the beginning of the winter vacation. Second, his family is together, his son and daughter home from their respective colleges. Third, he is driving home from the annual Christmas party.

Years ago, he resented having to work the day before Christmas. It was not the work he minded so much as the forced pleasantries at the annual party: the small talk, the cookies and punch, the miserly token "thanks" extended by management to the workers. However, as the mill has hired on more South American workers, the food and atmosphere at the annual party have become better. Gone is the watery punch; the workers bring in jars of *horchata* and heaping plates of foods my father relishes but has difficulty naming afterward. They even bring in a radio and dance on the floor of the factory. My father does not dance, but he allows himself a vicarious dancing in their presence, letting himself imagine what it is like to turn and strut among the band saws, grinders, and lathes.

He is driving home, then, feeling uncommon contentment. His worries will return in a few days: worries about whether he has wasted his life, his family's safety, the state of the world. But, for now, he is happy. I picture it this way: him smiling, the truck's heater ticking as it churns out hot air, a light snow blowing across the highway.

A car speeds past him on the left, threading through the thick lines of traffic. It hurtles toward the cars blocking its path in the left lane, then turns sharply into the right to avoid them, cutting off a red minivan. The van brakes to avoid the car, teeters for a second on two wheels, then tumbles, end over end, off the road and down into the ravine alongside the highway. My father pulls over to the side of the road and runs down the ravine to see if the driver is okay. At this point, I have trouble visualizing all that he saw. He later tells me that there was a woman in the van, that she was unconscious, bleeding from where her head struck the steering wheel. A crowd begins to form around the wreck. The woman awakes from her concussion but seems disoriented. She asks for her baby. And it is then that my father notices the hole in the windshield and, further down the ravine, an infant's car seat resting in the snow.

He later tells me about how surreal it was, picking his way down to base of the gorge, with the sun setting on the hills, to look for this dead child. The sheer stillness of those long moments is what he found most unsettling—the way time drew to a crawl, as if to offer plenty of opportunity to dwell on his anxiety, both the anxiety at the prospect of finding the child and the anxiety of having to carry that news back to the child's mother. Fortunately, it turns out that there is no dead child. Someone from the crowd calls down and tells my father to come back. Someone has found the woman's license, has called information to find her home phone number, and has called her home. Her husband answers; he explains that he has already picked up their child from daycare. The baby is safe at home.

There are more details to this story, such as the conversations that pass among the crowd as they wait for the ambulance, how the man who ran the woman off the road returned to see if she was okay but sped away. However, we can pass over these details. It is enough to mention one more moment. My father comes home. He embraces us: my sister and my mother and me. And he cries. And it takes him some time to explain why he is crying.

This is a face-to-face encounter. The moment when my father pulls over to the side of the road. The moment when he rushes to the wreck. The long drawn-out moments when he wanders along the ravine. And, perhaps most of all, the moments afterwards, when he is holding us and, over the next few days, when he keeps puzzling over why it has upset him so much. All of it is a face-to-face encounter. Perhaps even

I now, at this moment, writing these words, trying to understand what he experienced, trying, at the same time, to say something to you about the point of this book and why this moment lurches out of my memory when I think about what Levinas means by face-to-face encounters. This too is a face-to-face encounter.

3. What Is a Face-to-Face Encounter?

Levinas gives, of course, several examples of his own. Often, he asks the reader to imagine looking into the faces of persons who are helpless. We are asked to imagine passing by "the stranger, the widow, or the orphan" who is begging for help and feeling our hearts go out to them, feeling obligated to "tear the bread from our mouth to nourish the other." Other examples are not so dramatic. Levinas gives the example of holding a door open for a stranger as a mundane example of a face-to-face encounter. It is a simple act of civility, but Levinas is struck by the poetry of this gesture, the message it conveys about ceding one's place in the world to make room for others, about how easily such selflessness filters through our daily lives.

Two of my favorite examples come from interviews he gave in the eighties. At the time, Levinas was fascinated by Vasily Grossman's *Life and Fate*, a novel about the siege of Stalingrad and life in Stalinist Russia.[3] He cites two scenes from *Life and Fate* as illustrating face-to-face encounters. In one, Grossman describes a line of people standing outside of the notorious Lubyanka prison in Moscow. The line forms each morning, with hundreds of families waiting to make inquiries and to pass along packages and notes. A detail obsesses Levinas: Grossman describes how each person cranes their necks as they approach the prison's windows, hoping to glimpse their loved ones inside. Levinas dwells on this image, imagining each person among the hundreds staring at the neck of the person in front of him, each "reading on the nape of the person in front of him the feelings and hopes of his misery." This image of "the nape" can be understood, he says, as the "welcome of the face."[4]

In the other scene, a Russian woman encounters a captured German officer in the ruins of Stalingrad. The siege has been broken, and the officer is being forced to retrieve corpses from the rubble. A crowd gathers to watch him, among them a woman who has endured some great harm over the course of the war. The sight of the officer fills her with rage. She picks up a brick, moves to strike him down, but at the last moment throws down her stone. "Not understanding what was happening to her, governed by a power she had just now seemed to control," Grossman writes, "she felt in the pocket of her jacket for a piece of bread that had been given to her the evening before by a soldier. She held it out to the German officer and said: 'There, have something to eat.'"[5]

Elsewhere, particularly in later writings, Levinas's examples of face-to-face encounters are, arguably, less sanguine and more confrontational, involving elements of violence.[6] He associates the encounter with the other, what he comes to term "proximity," with situations involving what he variously calls "accusation," "persecution," "subjection," "obsession," "passivity," and being "held hostage" by the other. The other is not simply the colleague for whom I politely hold open a door, or the defenseless stranger to whom I beneficently offer help. The other is the "oppressed who is other than myself" who accuses me for my complacency and affluence.[7] The other is, for example, like the colonial subjects in John Coetzee's stories, the oppressed person who confronts her oppressor and accuses her for profiting from long-accepted injustices. Even more strikingly, the other can be the person who oppresses me, the victimizer who harms me yet for whom, Levinas claims, I am still "infinitely responsible."[8]

What do these examples tell us about the idea of face-to-face encounters? Levinas uses them, first, to link humanity with vulnerability. One of the terms he uses to talk about the other is *autrui*. *Autrui* corresponds to our use of the word *others* to refer to other persons. Some of Levinas's peers used it in discussing the "problem of other minds." For example, Sartre writes about solipsism and *le probleme de l'existence de l'autrui* in *Being and Nothingness*. Levinas distinguishes his use of this term, however, from its use in these debates. He insists that the other person is not fundamentally an "alter ego," another I, another mind or consciousness, and writes that "[t]he Other is the weak, the poor, the widow and the orphan, whereas I am the rich or the powerful."[9] He notes that "to recognize the other is to recognize a hunger."[10] So, Levinas uses *autrui*, in keeping with its standard use, to refer to other persons, but his examples of face-to-face encounters develop a suggestion about what situations most perspicuously reveal humanity. We are most keenly aware of other persons *as other persons* in situations where their vulnerability moves us to feel responsible for them. To recognize another person *as a person* is not to recognize a mind but a being that can be victimized, a being who depends on me for help.

Levinas's examples also illustrate a point about the force of the responsibility we feel for others. His examples of face-to-face encounters involve spontaneous acts of kindness. The woman in the second scene from *Life and Fate* cannot explain why she gives up her bread. She does not understand the "power" that moves in her, and her behavior seems foolish to her afterward. Similarly, holding a door open for a stranger is an act of kindness that we perform without thinking about it. Now, one might read Levinas as endorsing a kind of moral anarchy, suggesting that rules and norms get in the way of kindness and that we should engage in random acts of charity.[11] However, he is making a different point. What impresses him about these cases is not that they involve going *against* moral rules;[12] they inspire him because they show that ethics does not depend on rules for its authority. The woman who gives up her bread, despite her hatred, or who manages to show compassion even towards those who have harmed her, does not reason her way to this action. She finds it inexplicable, yet she still does it. Because her action is inexplicable, it shows something about the authority of moral responsibility. In a face-to-face encounter, one recognizes oneself as *unconditionally* responsible for the other, so responsible that one feels obligated to give *even though* one may not be able to rationalize it.

I can think of moments in my life that match these descriptions. There have been moments when I have seen strangers suffering and have been moved to help them. There have been more dramatic moments, too, of witnessing horrific acts and feeling as if my life was held hostage by them, as if it would be libelous to let myself forget or return too easily to day-to-day life. I think of watching news reports in 1994 of the genocide in Rwanda, the 9/11 terrorist attacks, or, most recently, watching documentaries on the genocide in Darfur and feeling shamed by my powerlessness and the world's reluctance to do anything to stop it. There also have been small moments of everyday kindnesses. However, for me, in my own life, it is the story my father told on December 24, 1999, that stands out as the best example of a face-to-face encounter I have experienced. Why?

Levinas associates the encounter with the other in places with "awakening" and "sobering up."[13] Levinas was fascinated throughout his life with wakefulness. He repeatedly uses it and insomnia as metaphors for consciousness and for human life in general, and they play a complex role in his work. Nonetheless, it is possible to skirt some of these details and to offer rough suggestions as to what he has in mind in associating face-to-face encounters with sobriety.

Levinas claims at one point in *Time and the Other* that our relationship to the other tends to be obscured in our everyday lives. It is "veiled by decency."[14] Typically, I am not aware of others *as others*. I live in a kind of fog composed of *my* beliefs, *my* desires, *my* ambitions, *my* life projects. I interact with others in this fog, but do not relate to them as unique individuals in their own right; I relate to them as if they were actors and actresses, *dramatis personae* in the story *I* am living, the unfolding story of *my* world. These interactions are structured, finally, by the social rules and etiquette of my tribe, my community's collective "veil of decency." Levinas's claim in *Time and the Other* is that there is something beyond these rules and this fog. His metaphor of a veil is crucial. Just as a face gives shape to a veil, so too the world of our everyday lives is shaped and given meaning by others. At the same time, this veil threatens to obscure this ground. There is a danger of getting lost in social rules and in our own personal life dramas, of forgetting what makes them meaningful.

A face-to-face encounter is a moment that disrupts this process: it is a moment when you wake up. The points I made earlier still stand: face-to-face encounters are moments when we recognize humanity as vulnerability, and when this realization impresses itself with unconditional force; but they are also moments when, so to speak, the fog parts and the gravity and solemnity of life become clear. It is for this reason that Levinas equates face-to-face encounters with sobering up. We tend to become drunk on our lives; in doing so, we forget what makes them meaningful. I become intoxicated with getting my book published, getting my students to succeed, fixing kinks in a romantic relationship, keeping in touch with old friends, seeing my party voted into power, getting bills paid, getting the peppers to grow in my garden. These dramas consume me. And then an event befalls me that both tears me out of them and clarifies them to me. I am sobered up. That is, I become aware that my life's dramas do not matter in their own right. There is something beyond them that matters more: the well-being of others. And it is their well-being that invests these dramas with meaning.

It is for this reason that my father's story stands out in my memories as an example of a face-to-face encounter. It is not simply that he came face-to-face with suffering or spontaneously rushed to help the woman. What interests me about this episode was its influence on him. It was a moment when he was sobered up. In a way, nothing changed in his life. His job remained unpleasant. The commute home remained long. He felt the same frustrations about where he had ended up in his life. He had been grateful to have his children home before the accident; he remained grateful afterward. Yet, in a way, everything changed. His frustrations appeared different, less overwhelming. They did not disappear, but they were brought into clearer perspective by having been moved by the loss the woman seemed to have undergone. They appeared as what they were: common frustrations and longings for respect and a few moments of leisurely peace.

The loves he had known prior to this episode were intensified by it as well. Here I find myself trying to imagine what it was like to wander at the base of that gorge. My father felt anxiety at the prospect of having to carry the news to the woman that her child was dead, but surely this anxiety was influenced by his awareness of how distraught he would be in that situation. The pity he felt for her led him to recognize something about the profundity of the love he felt in his own life, as well as its precariousness, how fragile it is, how easily threatened. Surely that is part of the reason why he embraced me and my sister so intensely, part of the meaning of this gesture. It was not that he was unaware of the fact that he loved us; one reason why he had been so content prior to the accident was because his children were home. However, his pity for the woman helped him to more fully realize what he already knew. In this regard, it was not unlike the pity Levinas associates with the first scene from Grossman. When Levinas visualizes Grossman's description of the Lubyanka prison, he pictures each person staring into the craned neck of the person before her, each empathizing with the next's misery, each gleaning, in turn, an insight into her own suffering and loves and, beyond them, into "all the anxiety in the world."[15]

4. Sobriety and Amnesia

I dwell on this example because it helps me to say something about this book's larger project. I said earlier that I am worried by the tendency in Levinas scholarship to overemphasize his epistemological claims.[16] I also said that this focus make sense since he does not advance many norms for regulating conduct. Typically, he points out moments when philosophers have tried, unsuccessfully, to conceptualize what he calls "the absolute other" (*l'absolument autre*). Nonetheless, I differ in how I interpret this point. The tendency is to read him as identifying a series of cognitive failures in Western philosophy. The concept of the absolutely other presents philosophers with an impossible cognitive challenge, and Levinas shows how they have ignored this fact. He repeatedly claims that each "human other" bears a trace of infinitude, transcendence, or "the absolutely other." The goal of philosophy, however, is to explain the totality of our experiences. Thus, it looks as if philosophy cannot help but misrepresent what is involved in acknowledging another person. For each person is an embodiment of the infinite, and philosophy cannot help but suppress this fact since it seeks to encapsulate all experiences within the strictures of its theories.

My belief is that this approach oversimplifies Levinas's worries about philosophy. Think again about my example of a face-to-face encounter. A face-to-face encounter is a moment that sobers me up, a moment when the responsibility I feel for another leads me to appreciate how there is something—the well-being of others—that is greater than my own personal window onto the world, and that invests this window with meaning. For my father, this sobering up occurred when his pity for the woman whose child seemed to have died led him to see something about the mundaneness of his dissatisfactions with life and the profundity of the loves he experienced in it. Yet these realizations fade. It took him days to recover. Yet a time came when his life's frustrations seemed overwhelming again and when the love he felt for others took on a formal, perfunctory cast.

This process of returning to our old ways may seem natural enough, but if we reflect on it, it should strike us as surprising, terrifying even. For consider what happens when we are sobered up by a face-to-face encounter. We do not discover a trivial insight. There is typically an element of shock and shame at oneself. How could I not see how insignificant my life problems are? Why did these far greater injustices not move me? How did I forget to love the people I love with the intensity they deserve? How could I not see how much depends on even this small act of acknowledgment which I give to this stranger? Face-to-face encounters are ennobling because they reveal the depth of our capacity to care for others, but they are terrifying because they reveal how infrequently we live up to this potential. You are not the person, you suddenly realize, you know you should be. You have not been this person for some time.

My suggestion is that Levinas's critique of philosophy is actually intended to address this crisis. He is typically read as pointing out some widespread cognitive failure in philosophy, but his true goal is to understand what I would call our amnesia of the good. He wants to clarify how we manage to lose sight of that sober perspective that opens up in face-to-face encounters.[17] How do we forget to give to others the support they deserve, given that we experience its urgency as all but crushing in these encounters? We shall see, too, that Levinas thinks that we can learn a great deal about the foundations of our everyday *Lebenswelt* by reflecting on the relationship to the other as it is evinced in face-to-face encounters. The other is, he thinks, the source of all language, thought, and agency. Recognizing the other requires us to acknowledge, finally, a being whose authority over us is absolute. To recognize the other is to recognize that you are "infinitely responsible" for her. Yet somehow, despite *all of this*, most of us go back to our daily lives. Our world is riven and what matters most is clarified. Yet we return to sleep. How?

Levinas's hunch is that there must be something about the concepts we use to reflect on our world that abets this amnesia. There must be something about the "veil of decency" that lets it be exploited in this way. As a philosopher, he turns to the ideas and concepts we inherit from the Western philosophic tradition to see where they permit us to falsify our relationship to the other. It seems to me that this is the real point of his critique of certain philosophies as "totalizing." It oversimplifies too much to depict him as drawing attention to a series of cognitive errors. His goal must be understood as fundamentally ethical—not to identify the limits of philosophy but to identify the conceptual mechanisms by which we allow quietism to take root in the world by misrepresenting our relationship to one another. His goal is to deprive us of these crutches so that we can linger in that moment of sobriety.

5. What I Try to Do in This Book

My main targets of criticism in this book, as I explain in chapter 1, are critics who argue that Levinas, for all his talk about ethics, cannot give us anything like a useful, practical ethics. If you have read Levinas, this criticism may be familiar. It draws on an observation I have mentioned several times. Levinas seems to think that each person is an embodiment of the infinite or transcendent. Ethical theories try to identify norms, rules, or principles we can use to explain and regulate human behavior. Yet if a human being *just is* what cannot be pinned down or systematized in this way, then any ethics Levinas could give would violate this teaching.

This book develops a defense of Levinas against this critique. Chapter 1 explains the criticism itself and points out one of its assumptions—namely, that it is impossible for Levinas to develop a practical ethics because he thinks that the "human other" (*autrui*) cannot be comprehended. To assess this inference, however, we need to determine why he thinks that the human other cannot be comprehended, why the other is supposed to constitute an embodiment of the infinite. It is one thing to say that we cannot comprehend the proposition *p and not p* being true; it is another to say that some of the cities in Italo Calvino's *Invisible Cities* cannot be comprehended. It is still another to point out that I cannot fully appreciate, from my perspective, what it is like for you, from yours, to taste a certain glass of chardonnay. It can be impossible to comprehend things for any number of reasons, and not all preclude or resist description or representation to the same extent or in the same ways.

Chapters 2 shows, in turn, that it is a mistake to think that Levinas's rationale for insisting that the other cannot be comprehended precludes him from developing a practical ethics. My argument for this thesis is complex, and much of my explanation will have to wait until chapter 2, where I analyze a pivotal section of *Totality and Infinity*. I draw on work in analytic metaethics, particularly writings by analytic philosophers on moral realism, to clarify Levinas's argument in this section, contending that he should be read as defending an extremely thoroughgoing form of moral realism. However, the gist of my argument is this. Levinas shows in *Totality and Infinity* that we cannot help but acknowledge others as having an unconditional claim on our care and support. He also notes that we tend to mischaracterize this relation by looking upon others as if they were objects of knowledge rather than responsibility. He uses terms such as "transcendence," "infinitude," and "absolute otherness" to stress how recognizing our responsibility for others compels us to make *practical* changes in our lives by engaging in *acts* of charity and justice. His point is not so much that the human other *cannot* be comprehended, but that it is a mistake, a dangerous one, to think of our relationship to her as one of knowing or cognition in the first place.

Here I should add as a warning that chapter 2 is not the only place where I draw on analytic philosophy to analyze Levinas. More than other commentators, I downplay Levinas's connectedness to figures and movements in continental philosophy, such as his place in phenomenology, his ties to deconstruction, and more recent French- and English-language continental scholarship on him. My intent is not to ignore this material, and I discuss figures such as Jacques Derrida, Jean-François Lyotard, and Michel Foucault in developing my defense of Levinas, as well as English-language commentators who have popularized interpretations of him that draw on these figures, such as Robert Bernasconi, Simon Critchley, and Jill Robbins. Nonetheless, I devote more attention to drawing Levinas into dialogue with trends and figures in analytic ethics—such as debates in analytic metaethics (chapter 2), Rawls's theory of justice (chapter 4), and the revival of pragmatism (chapter 5). I choose this focus because it seems to me that more needs to be done to situate Levinas in this tradition. There is a massive secondary literature on Levinas's place in continental philosophy, yet relatively few have sought to bring him into dialogue with debates in analytic philosophy.[18] My hope is that reflecting on Levinas from this new vantage point—assessing his claims about ethics in light of, say, Rawls's theory of justice or debates over moral realism—may shed light on features of his thinking that have gone underappreciated, if only because of the heavy focus in continental philosophy on issues of language, antifoundationalism, and the limits of expression.

Chapters 3 and 4 offer a more straightforward defense of Levinas. Levinas's critics fault him for not offering anything like a practical ethics. I show that this criticism is misplaced by demonstrating that he does, in fact, indicate what norms should regulate conduct. Chapter 3 surveys political writings where Levinas endorses Kantian and liberal ideals. It also takes up the question of why he does not offer more explicit advice. Levinas is reluctant to advance a full normative ethics, I argue, because a pragmatic sensibility infuses his work. He sees ethical theories as tools we use to negotiate the challenges we face in balancing our responsibilities, and he is reluctant to advance a single ethics because flexibility may be needed to discern which will prove most effective in resolving a given dilemma. As a first stab at indicating his ethics, then, I argue that Levinas begins with a broadly Kantian, liberal outlook. He defends this starting point on pragmatic grounds. His claim is not that Kantian liberalism is, in fact, *true*, but that, for a variety of reasons, it constitutes the best set of "tools" available to us.

Chapter 4 refines this suggestion by assessing the extent of Levinas's compatibility with sophisticated forms of Kantian liberalism of the type espoused by Jürgen Habermas and John Rawls. Some commentators, such as Simon Critchley and Steven Hendley, have argued that Levinas's claims about ethics should be wedded to these sophisticated forms of Kantianism. I show that Levinas would be sympathetic to critiques developed by feminist philosophers and philosophers working in disability studies, and he would follow them in arguing that even these sophisticated forms of Kantianism are flawed in their failure to appreciate the moral significance of caregiving. Chapter 4 challenges the claim that Levinas cannot give an account of justice, then, on two levels. It is wrong because there is evidence that he would accept a more robust account of distributive justice than that found in Kantian liberalism. It also challenges this criticism by modeling how Levinassian modifications can be made to normative ethical theories. It seeks to show that it is wrong to fault Levinas's philosophy for being unworkable by modeling how concrete revisions to Kantianism can be developed based on Levinas's remarks on the relation to the other.

In a way, though, this entire defense of Levinas is a bit of a dodge. I initially set out to defend him against what I took to be the most significant criticism of him in the secondary literature—namely, that he cannot give his readers anything like a practical ethics. In defending him, however, I found myself weighing in on a larger issue. For this criticism hinges on a certain understanding of why Levinas thinks that the other cannot be comprehended and on how two sides of his philosophy, the practical and the theoretical, fit together. My concluding chapter, chapter 5, takes stock of the overall approach to Levinas implied by my defense. I characterize this approach as pragmatic in that it reads Levinas as primarily troubled, like pragmatists, by our tendency to adopt detached, spectatorial perspectives on the world and our lives and as concerned to root out this tendency so that we will adhere to a standpoint of engaged agency. Chapter 5 reviews some of the basic interpretations of Levinas in the secondary literature and assesses how this pragmatic approach fairs against them. In particular, I make a case for it against more deconstructionist readings, arguing against them that Levinas does not deny that there can be legitimate theoretical descriptions of the world and our lives, but claims that we should assess their legitimacy based on how effective they are in helping us to balance our responsibilities, whether they express our most creative solutions to the particular moral crises that beset us.

Yet even in summarizing my approach as pragmatic, I worry that I am not doing justice to the point I wish to make about Levinas. For while it is one thing to *say* that he hopes to disabuse his readers of detached perspectives, it is another to convey in a tangible way what this means. Thinking about the moment in my father's life when he embraced me, what was conveyed in his embrace, helps me to at least gesture at the larger point I hope to express through this book's conclusion. This book sets out to show that Levinas can and does supply his readers with the rudiments of a practical ethics. At the same time, I do not find it contradictory to admit that outlining such an ethics was not his top priority. Nothing in his work prevents him, *or should prevent us*, from reflecting on which ethical theories best embody respect for the other. Yet his more urgent goal was, I think, to *move* his readers so that they experienced something like my father's state of mind on the night he returned home from the accident. He hoped to inspire a kind of change in them, to wake them up. This book is an attempt to do so as well. It defends Levinas against a certain line of critique, and it uses this defense to root out various misunderstandings of him. However, rooting out these confusions is valuable only insofar as they currently serve to soften the practical demands that Levinas's philosophy makes on us. At bottom, then, this book is perhaps best thought of as an attempt to let Levinas wake us up.

Endnotes

1. See Levinas, "Transcendence and Evil," in *Collected Philosophical Papers*, trans. Alphonso Lingis (Pittsburgh: Duquesne University Press, 1987), 175–186. *See esp.* p. 182 for his discussion on how evil produces a "de-neutralization of being."
2. Critchley, "Introduction," *Cambridge Companion to Levinas*, ed. Simon Critchley and Robert Bernasconi (Cambridge: Cambridge University Press, 2002), 27.
3. Grossman completed *Life and Fate* in 1960, but it was immediately confiscated by the KGB and was not published until the early eighties, after it was smuggled out of the Soviet Union. Levinas frequently cites the novel in interviews he gave in the eighties as exemplifying his philosophic outlook.
4. Levinas, "In the Name of the Other," in *Is it Righteous to Be*?, ed. Jill Robbins (Stanford: Stanford University Press, 2001), 188–199. *See esp.* p. 191. See also Levinas, "The Other, Utopia, and Justice," in *Entre Nous*, trans. Michael B. Smith and Barbara Harshav (New York: Columbia University Press, 1998), 223–233. *See esp.* pp. 231–232.
5. Grossman, *Life and Fate*, trans. Robert Chandler (New York: Harper and Row, 1985), 805.
6. Here I should acknowledge a worry one may have about my presentation of Levinas in this Prologue, and perhaps more generally in this book. One might worry that I am romanticizing Levinas in neglecting to stress the violent, confrontational elements of the encounter with the other. To be clear, my intent is not to ignore the ways in which the encounter with the other involves violence, destabilization, or confrontation, and it is worth noting that Levinas stresses these associations in later writings, particularly *Otherwise than Being*.

   Nor is my intent to romanticize Levinas elsewhere in this book. Simon Critchley has pointed out that there is tendency among Levinas scholars to lapse into "homage" and "discipleship"—to affect a certain breathlessness in discussing Levinas and to uncritically accept his ideas. I am also troubled by this tendency. While I defend Levinas, I hope to avoid the trap of devolving into hagiography. I do not dwell on moments where he makes claims on matters of ethical or political significance that trouble me only because my aim is to defend him against critics who allege that he has *nothing valuable* to say on such matters *whatsoever*. I do think that there are areas in his philosophy where we should question him—such as, for example, his characterizations of femininity. See Critchley, "Five Problems in Levinas's View of Politics and the Sketch of a Solution to Them," in *Political Theory* 32, no. 2. for more on Critchley's concerns about the state of Levinas scholarship. For feminist analyses of Levinas, see Chanter's *Ethics of Eros: Irigaray's Re-reading of the Philosophers* (London: Routledge, 1995) as well the collection she edited, *Feminist Interpretations of Emmanuel Levinas* (University Park, PA: Pennsylvania State University Press, 2001). I am indebted to an anonymous reader for alerting me to the worry that I may be romanticizing Levinas.

7. Levinas, *Otherwise than Being*, trans. Alphonso Lingis (Pittsburgh: Duquesne University Press, 1998), 55.
8. See for example Levinas's *Ethics and Infinity*, trans. Richard Cohen (Pittsburgh: Duquesne University Press, 1985), 99–100.
9. Levinas, *Time and the Other*, trans. Richard Cohen (Pittsburgh: Duquesne University Press, 1987), 83.
10. Levinas, *Totality and Infinity*, trans. Alphonso Lingis (Pittsburgh: Duquesne University Press, 1969), 75.
11. Elements of Zygmunt Bauman's *Postmodern Ethics* (Oxford: Wiley-Blackwell, 1993) are suggestive of this type of interpretation of Levinas. See also some of Simon Critchley's remarks in his more recent writings on Levinas, such as "Five Problems in Levinas's View of Politics," *Political Theory*, and *Infinitely Demanding: Ethics of Commitment, Politics of Resistance* (New York: Verso, 2007).
12. For example, it is hard to imagine Levinas showering the second scene from *Life and Fate* with praise if the woman knew she was violating a moral rule—if, say, she knew the officer was plotting to murder an innocent and if a few pages later he did so because he was finally strong enough, having eaten the bread, to perform this task.
13. See, for example, Levinas "Philosophy and Awakening" in *Entre Nous*, trans. Michael B. Smith and Barbara Harshav (New York: Columbia University Press, 1998), 77–90. *See esp.* pp. 87, 89; Levinas "Philosophy Justice and Love" in *Entre Nous*, trans. Michael B. Smith and Barbara Harshav (New York: Columbia University Press, 1988), 103–121. *See esp.* p. 114.
14. Levinas *Time and the Other*, 82.
15. Levinas, "In the Name of the Other," in *Is it Righteous to Be*?, ed. Jill Robbins (Stanford: Stanford University Press, 2001), 188–199. *See esp.* p. 192.
16. The *locus classicus* of this tendency is Jacques Derrida's "Violence and Metaphysics," in *Writing and Difference*, trans. Alan Bass (Chicago: University of Chicago Press, 1978), 79–153, which popularized *Totality and Infinity*

and set the tone for much of the subsequent commentary on it. Derrida canvases various points in *Totality and Infinity* where Levinas seems to commit the same "totalizing" errors which he faults in Hegel, Husserl, and Heidegger, and Derrida more generally raises the question of whether Levinas respects the limits he himself sets with regard to the comprehensibility and expressibility of the other.

17. Levinas's interest in this crisis is understandable, given his biography. Levinas was not himself a Holocaust survivor. However, he did witness firsthand the cruelties perpetuated against the Jews by the Nazis in that he spent the bulk of World War II in an intensive Nazi labor camp for Jewish prisoners of war. Moreover, his entire family, all Lithuanian Jews, were murdered over the course of the war. Finally, his early philosophic mentor was Martin Heidegger, who, infamously, was the leading philosophic spokesperson for the National Socialist Party in Germany in the early thirties. It is understandable, then, that Levinas would be preoccupied with the question of how people manage to forget the good and how various conceptual systems lend themselves to facilitating this amnesia.
18. Michael Morgan's *Discovering Levinas* (Cambridge: Cambridge University Press, 2007) is perhaps the most notable effort to bring Levinas into conversation with Anglo-American philosophy. Prior to Morgan, a handful of prominent figures in analytic philosophy had begun to take an interest in Levinas, such as Hilary Putnam and Stanley Cavell. Their writings on him, though, tended to be limited reflections on how elements of his thinking resemble ideas in their own work. Morgan's book is the first to develop a comprehensive interpretation of where Levinas stands in relation to analytic philosophy. Simon Critchley notes that an effort was also made in *The Cambridge Companion to Levinas* to achieve a "balance between the more usual phenomenological or Continental approaches to Levinas's work and more analytic approaches." See Putnam's "Levinas and Judaism," 33–62 and Bernstein's "Evil and the Temptation of Theodicy," 252–267 in *The Cambridge Companion* for examples of approaches that are more analytic.

Emmanuel Levinas on the Priority of Ethics

Chapter 1

Why Is It Hard to Talk about Justice?

And I always thought that the simplest words
Must be enough. That when I say how things are
Everyone's heart must be torn to shreds.

—Brecht, from his last poem

1. Introduction

My topic in this book is Emmanuel Levinas's philosophy. Levinas is increasingly coming to be regarded as one of the most important figures in twentieth-century European philosophy. He was instrumental in introducing Husserl's and Heidegger's work to French philosophers, and he was an influential figure both in existentialism and in more recent postmodern philosophy. He is best known for his powerful claims about ethics. Levinas is often associated with his slogan that "ethics is first philosophy," an enigmatic thesis, but one that conveys something of how important he finds ethical considerations to be for philosophy and for our lives generally. His writings are studded with moving descriptions of what it is like to look into the faces of those who suffer, and he is fascinated with the pity such moments evoke. Indeed, perhaps *the* central aim of his work is to expose the extent to which our lives and thinking are defined by the aptitude for selflessness, what he sometimes calls "saintliness" or "holiness," which confrontations with human vulnerability reveal in us.

My goal in this book is to defend Levinas against what I believe to be the most serious criticism of him in the secondary literature. This criticism can be put this way: Should we not be troubled by the paucity of advice he gives on how to be ethical? Does it not reflect poorly on his claims about ethics that he never recommends standards for determining ethical conduct? I want to defend Levinas against this criticism. I think he has useful advice about how to be ethical; his critics simply have not read him attentively enough, or, better, they have been so convinced that he *cannot* offer substantive suggestions on how to determine ethical conduct that they have overlooked moments when he offers advice on this score. Moreover, he has interesting reasons for not offering more advice. For example, we will see that Levinas considers normative ethical theories to be tools we use to attune ourselves to one another's needs and to balance our "infinite responsibilities" to multiple "others," and he is reluctant to advance any single ethics because flexibility is needed to discern which ethics best enables us, in a given situation, to render ourselves maximally responsive to one another. His reticence as to what norms should guide ethical conduct is not indicative of skepticism on his part as to whether they can be generated but of a hardheaded pragmatism about the real-world challenges involved in using them effectively.

My more immediate task in this chapter, however, is simply to clarify the basic criticism leveled against Levinas. I say "clarify" because I think the criticism as it is stated is misleading. Levinas's critics fault him for not outlining anything like a normative ethics, but their real concern is not that he *does not* but that he *cannot* offer his readers anything like a normative ethics. Their worry is that he leaves his adherents in a double bind.

On the one hand, he stresses the need to be ethical. On the other hand, he depicts this demand in such a way that he cannot specify any norms for determining ethical conduct. Their further worry is that this tension is unbearable; we should mistrust a theory of ethics that makes it impossible to specify any ethical norms whatsoever. My ultimate goal, again, is to show that Levinas offers suggestions as to what norms or ideals should guide ethical conduct. Yet much of this book is devoted to a more modest claim: I want to show that he *can* say substantive things on this issue. My goal in this chapter is to begin to motivate this approach by clarifying the charges against Levinas.

2. Is Levinas a Nuisance?

Several philosophers have recently argued that Levinas's philosophy is marred by his failure to develop useful, action-guiding ethical principles or procedures. Levinas talks a lot about ethical responsibility and the role ethical considerations should play in philosophy. He claims that "ethics is first philosophy," and he repeatedly alleges that in each of our lives, we undergo exceptional moments, face-to-face encounters with the other, in which we come to realize that we are "infinitely responsible" for each and every person. Yet he never indicates how to put his view into practice; he never indicates what normative ethics would best express his analysis of the relation to the other. Should we not be troubled by this reticence? Does it not reflect poorly on his claims about ethics that he never recommends policies for determining ethical conduct? Does it not say something about the infeasibility of his ethic of "unlimited care"[1] that he never indicates how to practice it?

Richard Rorty has given the sternest formulation of this criticism: "The notion of 'infinite responsibility' formulated by Emmanuel Levinas," he observes, "may be useful to us in our individual quests for private perfection. When we take up our public personalities, however, the infinite and the unrepresentable are merely nuisances."[2] Reading Levinas may be personally rewarding, but his work is useless, a "nuisance," when it comes to the practical task of proposing concrete policies and institutions for regulating civic conduct. Simon Critchley expresses similar worries. Critchley argues that Levinas's work is not without merit.[3] Nonetheless, we live in a world in which we are faced with urgent yet complex ethical dilemmas; it is reasonable to expect some guidance from Levinas on these matters. Yet he never offers any advice: he leaves our thinking "fatally undetermined."[4] "Levinas tells us that his ethics must lead to some theory of justice," Critchley observes, "without telling us in any detail what this theory might be."[5] Elsewhere, Critchley suggests that the gap between Levinas's characterization of ethical responsibility and his unspecified theory of justice may represent "the fundamental *aporia* of deconstruction."[6] And in recently published essays, he faults Levinas's failure to specify his theory of justice as "the Achilles' heel of his work."[7]

It is worth pausing to reflect on these criticisms. Rorty and Critchley make it sound as if they are troubled by Levinas because we live in especially perilous times. We find ourselves confronted with ethical dilemmas that it would be irresponsible to avoid. Levinas *talks* a lot about "infinite responsibility," but his philosophy does not really help us confront these pressing dilemmas. Thus it should be rejected. Yet this makes it sound as if Rorty's and Critchley's view is that all philosophers should be ethicists or, better, activists: the only good philosopher is an engaged one. This may be Rorty's view, but it does not seem like a claim Critchley would endorse.[8] Nor does this objection identify a flaw specific to Levinas. For if the argument is that we should reject *any* philosophy that is not directly applicable to contemporary ethical and political dilemmas, then this objection can be raised against any philosopher who is not primarily concerned to outline a concrete normative ethics—to anyone working in logic, metaphysics, epistemology, or the history of philosophy.

Perhaps the complaint is that it is misleading for Levinas to depict his work *as ethics*, given the paucity of advice he gives on how to determine ethical conduct. Critchley invites this response when he asks if "ethics is the right word to describe the basic experience that [Levinas] is trying to convey."[9] The problem with Levinas, on this reading, is that all his talk about "ethics as first philosophy" is apt to make us think that he is trying to do ethics, when he offers something quite different. Along these lines, Critchley distinguishes between two types of moral philosophers in discussing Levinas: "moral perfectionists" and "legislators."[10] A perfectionist argues that "ethics has to be based on some form of basic existential commitment…that goes beyond the theoretical strictures of any account of justice or socially instituted ethical code." She believes that "an ethical theory that does not give expression to this basic demand will spin in a void and have no compelling way of explaining the source of one's motivation to act on the basis of that theory." By contrast, a legislator is interested in identifying "detailed precepts, rules, and principles that add up to a theory of justice." Critchley argues that Levinas is a perfectionist. He is interested in something closer to what we might call metaethics; he wants to show us something about the nature of ethical responsibility itself, the nature of the ethical demand. He leaves it to other philosophers to take up the task of spelling out what is required to be ethical.

Yet how much concrete advice does a philosopher need to give for her work to qualify as ethics? And how firm is this distinction between perfectionists and legislators? Consider Kant. Kant's *Metaphysics of Morals* is a relatively understudied text; philosophers interested in his ethics tend to focus on *Groundwork of the Metaphysics of Morals* and *Critique of Practical Reason*. Kant's goal in the *Metaphysics of Morals* is to explain how his ideas about duty and respect should be put into practice. However, it is not clear that he has this ambition in earlier works. He indicates in *Groundwork* how the categorical imperative can be used to gauge the permissibility of various forms of conduct, and he offers suggestions in it and in the second *Critique* about how earlier ethical theories erred in identifying morality with the pursuit of happiness, but his main aim is not to supply us with "rules, principles, and precepts that add up to a theory of justice." His aim is to explicate what he calls "the laws of freedom,"[11] the principles that underwrite our conceptions of ourselves as agents. So, what is Kant—a perfectionist or a legislator? His aim in the *Groundwork* and second *Critique* seems to be perfectionist in its focus on the foundations of ethics. Yet it seems counterintuitive to say that he was not really doing ethics until he wrote the *Metaphysics of Morals*, or that there is not material in these works that is of significance for moral legislators. Nor is Kant the only ethicist about whom this point can be made; G. E. Moore's *Principia Ethica* and Thomas Nagel's *The Possibility of Altruism* do not bristle with advice on euthanasia, censorship, or distributive justice. Yet we do not flinch when they are cited as important works in twentieth-century moral philosophy.

So, why not view Levinas as similar to these authors? His core ambition may not be to specify a comprehensive normative ethics, but he does reflect on the nature of ethical responsibility and the underlying concepts that define our view of ourselves as moral agents.[12] He also indicates where previous ethical theories have erred in discussing these matters. For example, he faults social contract theories that view morality solely in terms of uniform social agreements between competing, self-interested individuals.[13] He also criticizes figures such as Hegel, Marx, Heidegger, and Althusser for privileging collective entities like states, cultures, nations, or classes over individuals as the primary datum of ethics and political philosophy.[14] So, again, why not think of Levinas as having a similar approach to ethics as Kant's in the *Critique of Practical Reason*? Kant's main aim may not be to advance a detailed normative ethics, but his texts are valuable for what they say about the nature of ethical responsibility, and there is indication in them of how we might generate ethical standards based on Kant's analysis. Similarly, why not say that Levinas's main aim is not to develop a normative ethics, but that nothing prevents us from developing one based on his remarks on ethical responsibility?

So, the criticism of Levinas can sound puzzling. I take it that Critchley's basic claim is that we expect something from philosophers working in ethics that we do not find in Levinas. Yet when we compare Levinas to other ethicists, it is not clear what is missing. Certainly one finds *more* indication in Kant as to what ethical policies he favors, and so my comparison of Levinas and Kant may seem misleading. However, my point is that the differences between them are quantitative rather than qualitative. Both are interested in clarifying the nature of ethical responsibility and what a moral agent must be like for ethics to be possible. Kant simply indicates in greater detail what ethical rules are implied by his discussion of these topics. Moreover, why should we think that Levinas's failure to specify a normative ethics *itself* constitutes grounds for dismissing his ideas? It may be *disappointing* that he gives little indication of how to put ideas into practice.[15] However, why should we think that this reticence alone entails that his claims about "the face-to-face encounter" or "totalizing" philosophies must be wrong? Again, it seems like an odd criticism, as if we could dismiss, say, Kant's claim that a "good will" is the only thing that deserves to be called "unqualifiedly good," or Moore's thesis that "good" is a simple, nonnatural, indefinable quality simply because they do not say enough in the *Groundwork* or in *Principia Ethica* about capital punishment or just-war theory.

I stress the ambiguity of the charge against Levinas because I think that it is misleadingly worded. The criticism sounds clear cut: it reflects poorly on Levinas that he only gestures at how to put his ideas into practice. However, the real criticism that philosophers like Rorty and Critchley intend to make is not that he *does not* say enough about how to put his ideas into practice but that, in principle, *he cannot say anything at all*. Indeed, it seems to me that Levinas's critics and defenders agree on this point: both believe it would go against the drift of his thinking for him to advance any sort of normative ethics. To justify this claim, however, I need to review some of Levinas's core theses. In particular, we need to know more about his ideas concerning the absolute transcendence of what he calls "the human other."

3. A Brief Introduction to Levinas: Transcendence and Humanity

I do not want to delve too far at this point into the details of Levinas's philosophy. Levinas's prose is notoriously obscure, studded with quasi-technical terms and rhapsodic digressions that are evocative but hard to explicate, and I fear I will lose sight of my main thesis if I try to explain too much of his thought in one sitting. For now, then, I just want to offer a sketch so we can appreciate why his critics have found it tempting to think that he cannot specify any normative ethics at all.

One of Levinas's main goals was to assess Heidegger's philosophy. Levinas was one of Heidegger's students, and he felt throughout his life that Heidegger was one of the great minds in the history of philosophy. He concluded early on, though, that there had to be a flaw somewhere in Heidegger's thought, and he came to think that Heidegger's philosophy devolved into an objectionable form of monism.

His criticism on this point owes a great deal to critiques of Hegel by philosophers such as Søren Kierkegaard and Franz Rosenzweig.[16] Kierkegaard and Rosenzweig were troubled by Hegel's conviction that all reality could be rendered wholly transparent to human reason. They were especially vexed by this view's ontological assumptions. If all reality is transparent to reason, it follows that all reality is reducible to a series of rational processes. As Hegel put it, "The Real is rational, and the rational is real."[17] Philosophers like Kierkegaard and Rosenzweig found this view galling. One of their worries was that Hegel's view leads us to disregard the richness, spontaneity, and utter particularity of human existence. Hegel invites us to regard all of reality as a kind of outward manifestation of an elaborate philosophic system, including our own lives. He invites us to perceive our lives not as unplanned, spontaneous, or utterly unique—things in which we are personally and passionately involved—but as akin to abstractions within a larger conceptual "totality."

Levinas was struck by this critique of Hegel. He was intrigued by its claim that Hegel's system is flawed because it reduces reality to nothing more than an outward manifestation of a single principle, *Geist*, or Mind. He came to think that Heidegger's philosophy devolved into a similarly oppressive monism. One of Heidegger's achievements in *Being and Time*, of course, was to stress the importance of prereflective life experiences, such as moods or habits, for philosophy. So, it may seem odd to fault him for espousing a monism that reduces all reality to a manifestation of *Geist* or Mind. However, Heidegger's distance from

Hegel is ultimately not that great. Being or *Sein* simply replaces *Geist* as the monolithic principle that threatens to distract us from the particularity of human existence.

Levinas's critique on this point takes two main forms. He sometimes argues that Heidegger's concept of truth as Being's "revelation" (*alētheia*) commits him to a similar thesis about the coequality of reality and rationality that one finds in Hegel and Parmenides. Heidegger may denounce much of Western philosophy as blind to the significance of prereflective life experiences, but he ultimately upholds a similar view insofar as he characterizes reality, or Being, as something that reveals or "discloses" itself in human consciousness. Thus Heidegger is susceptible to criticisms like those which Kierkegaard and Rosenzweig leveled against Hegel.[18] Elsewhere, Levinas focuses on Heidegger's phenomenological analyses of various aspects of everyday human existence and tries to show that, despite their richness, they depict it as nothing more than the successive manifestation of a single substance: Being.

Again, the details of these critiques need not concern us now. The key point to note is the more general worry about philosophic inquiry that Levinas develops in criticizing Heidegger. Levinas came to think that the errors that tainted Heidegger's thought pervaded much of the history of philosophy: most philosophers' theories committed them to the same reductive monism found in Hegel and Heidegger. He borrows a pair of terms from Plato's *Sophist* to discuss this monism. Plato argues in the *Sophist* that the categories of "the same" (*to auton*) and "other" (*to heteron*) are basic categories that cannot be derived from each other or any other categories, such as being or nonbeing.[19] Levinas invokes this distinction in his work, claiming that most philosophies privilege the category of "the Same" (*la Même*) over that of "the Other" (*l'Autre*). What he means by these terms is difficult to explain. It helps to note that he uses the French word *autre*, which corresponds to our use of *other* to refer to something distinct from things mentioned or implied. One would use *autre*, for example, to say *I prefer the other film. (Je prefere l'autre film.)* So, Levinas invites us, apropos of Plato's *Sophist*, to think of "otherness" as a foundational ontological/epistemic category. It is his term for both the antithesis of *haeceeity*, for "not-this-ness," and for what is conceptually unfamiliar, what is "other" than what is known or comprehensible. Thus he uses *autre* alongside such terms as "Wholly Other" (*Tout Autre*), "the absolutely other," "the ungraspable," "mystery," "enigma," "infinity," and "the transcendent."

Levinas proposes, then, that philosophic systems traffic in two basic categories—*la Même* and *l'Autre*, the Same and the Other, the familiar and the transcendent—and he refers to his own work as a "dualism" or "pluralism" that parses reality not into Mind and Matter, or Self and World, but into the familiar and the transcendent.[20] His corresponding worry is that most philosophies simply disregard the category of the transcendent altogether. Hegel is emblematic of this denial in his insistence that all reality can be rendered transparent to reason, but Levinas thinks most philosophic systems share this flaw. Philosophers are concerned, after all, to extend and refine our understanding of the world. So, the very enterprise of philosophy seems to be founded on the belief that the unfamiliar can always be made familiar. In Levinas's terms, it is founded on the perpetual assimilation of the Other to the Same. Thus Levinas thinks most philosophic systems are "allergic" to the very idea of anything being genuinely transcendent or "absolutely other."[21] This inability to countenance the possibility of genuine transcendence condemns most philosophies to the monism found in Hegel and Heidegger. Most philosophies devolve into the "imperialism of the same."[22]

One further thesis in Levinas needs to be reviewed before I can explain why his critics have assumed that he cannot advance a normative ethics. Levinas worries that most philosophies devolve into a reductive monism premised on a denial of the category of the "absolutely other." This would seem to suggest that he thinks that there are moments in our lives when we brush up against transcendence. That is, the problem with most philosophies is not simply that they are monistic, but that monism is, in fact, incorrect or reductive, and their acceptance blinds us to moments in our lives that defy comprehension. So, what does Levinas deem to constitute one of these transcendent experiences?

His answer is novel: we experience transcendence in our interactions with other persons. "The human other," as he puts it, "is absolutely other": "*L'absolument Autre, c'est Autrui.*"[23] Levinas's claim is that there is an important sense in which *other persons* defy our ability to comprehend them. In particular, there is something about witnessing another person in need of help—the shame, pity, and responsibility it invokes—that involves a premonition of transcendence.

This equation of humanity and transcendence, *l'autrui* and *l'autre*, helps explain why Levinas finds the monism characteristic of most philosophies so upsetting. Their error is not purely intellectual; it has ethical ramifications. The problem with most philosophers, that is, is not simply that they presume that all reality is reducible to one basic principle; the deeper problem is that this bias leads them to underestimate the uniqueness and particularity of human existence. To borrow Shakespeare's line, the problem is not simply that philosophers have trouble admitting that there is more in between heaven and earth than can be dreamt of in their philosophies; the deeper problem is that other persons are the chief thing that exceeds comprehension, and thus the insistence in philosophy on reducing all reality to something akin to the outward manifestation of a single, self-contained conceptual system blinds us to what is distinctive and exceptional about human nature.[24]

4. The Ineffability Hypothesis

I suggested at the end of §2 of this chapter that Levinas's critics and defenders alike assume that he cannot specify any procedures for determining ethical conduct without violating the central claims of his philosophy.

It should be clear why this view is tempting. Levinas's big idea is that we uncover something transcendent in our encounters with other persons; there is something about accepting responsibility for another person that involves a premonition of transcendence. However, if this is the case, then it seems impossible to specify any ethical norms or standards that are expressive of this experience of humanity. Let me refer to this explanation of why Levinas does not advance a normative ethics as the "ineffability hypothesis." The ineffability hypothesis makes the following argument. Levinas frequently writes about "the face-to-face encounter with the other." He uses this expression to refer to situations in which we become keenly aware of other persons as ineffable. In particular, he argues that this revelation occurs in the pity evoked in us by the suffering of others. If we scrutinize our reactions to human vulnerability, we will notice that they evince a premonition of others as absolutely transcendent. However, it follows that Levinas cannot specify any ethical standards that claim to be expressive of this premonition, for any claim to have accurately expressed it will contradict the claim that it is ineffable. His philosophy seems to lead, therefore, to a kind of "moral mysticism" in which the basis of morality stands beyond all intellectual apprehension.

The criticism discussed earlier makes more sense if we read it as relying on this hypothesis. Levinas's critics fault him for not saying enough about how to put his ideas into practice, but, as we saw, it is difficult to see how being reticent alone can constitute grounds for rejecting his entire philosophy. There may be grounds for rejecting it, though, if it makes it impossible to generate *any* ethical standards or renders *all* standards fundamentally controvertible. Again, we live in a world in which we are faced with urgent yet complex ethical dilemmas—dilemmas it would be irresponsible to avoid. Yet Levinas would seem to have us believe that there can be no nonmisleading ethical standards. One might worry that his philosophy leads to quietism. If we accept it, we may feel driven to give up on morality since any moral judgments we make will rely on arbitrarily chosen norms. Or we might worry that he cannot handle cases where people disagree about what is required by ethics. After all, if all ethical norms are misleading, then we need never accept others' judgments when they fault us for having false ethical views. Levinas's "moral mysticism" can seem to lead, then, either to quietism or to moral anarchy. Neither option is acceptable; hence his work should be rejected. At the very least, he seems to leave his adherents in a double bind. He stresses the importance of ethics yet depicts the ethical demand in such a way that it is impossible to determine how to act on it.

5. The Pros and Cons of Silence

Levinas was not oblivious to these concerns, and he sometimes draws a distinction in his work between "ethics" and "justice" that seems intended to address them. He uses "ethics" to refer to the face-to-face encounter with the other. I have already referred to the notion of "the face-to-face encounter" a few times in this chapter and in the Prologue. As we have seen, Levinas uses it to refer to situations where we are struck by another person's transcendence. He also uses it to refer to situations that vividly reveal to us what is involved in recognizing other persons as morally invaluable. His examples usually involve cases where one witnesses the vulnerability of others. A face-to-face encounter is a situation that jolts us out of complacency—that shocks us and forces us to recognize the preciousness of another person's life.

Levinas also writes about what he calls "justice" (*le droit*). He uses "justice" to refer to the gamut of things we typically associate with normative ethical theories and with social and political philosophy. A theory of justice is any system of norms, standards, or procedures used to assess the moral acceptability of our actions, social policies, legal institutions, and so forth. Levinas claims, finally, that justice flows out of ethics. We can think of our moral codes, social polices, and so forth as things we develop on the basis of the primordial ethical experience that occurs in the encounter with the other. They represent our best estimations as to what ways of life best express the respect we apprehend in such encounters. In particular, they represent guidelines we have developed in order to assist ourselves in the difficult task of adjudicating among competing claims on our time and attention—our competing responsibilities to multiple "others."

Much more can be said about the relationship between ethics and justice in Levinas. For now, though, I want to skirt this issue. The important point for our purposes is Levinas's apparent claim that we *can* develop ethical standards that are expressive of the face-to-face encounter with the other. Indeed, he seems to think it is not only *possible* but *necessary* to develop them. So, it seems that what I am calling the "ineffability hypothesis" cannot be accurate. It alleges that Levinas does not develop a normative ethics because he cannot do so without misrepresenting the awareness of other persons as "absolutely other" that occurs in the face-to-face encounter, but he explicitly denies this claim in distinguishing between ethics and justice.

Levinas's critics are not unaware of this distinction; they simply second-guess it. Consider again the passage from Critchley quoted earlier: "Levinas tells us that his ethics must lead to some theory of justice, without telling us in any detail what this theory might be."[25] Levinas admittedly *says* that his remarks on ethics should culminate in an account of justice, but he never fulfills this promise. Critchley's suggestion is that this failure is telling: it indicates that he cannot effect this shift from ethics to justice. He would need to subject interpersonal relations to exposition to develop a normative ethics, but he cannot consistently perform this act since, by hypothesis, they bring us into relation with something ineffable. More guardedly, he cannot make any *positive* claims about which principles of justice express the primordial experience of ethical responsibility that occurs in the encounter with the other. He can perhaps develop critical strategies that amount to the ethical equivalent of a negative theology. That is, he can fault existing ethical, legal, and political institutions as inevitably misrepresenting what it is like to acknowledge another person's humanity. He can, as it were, "deconstruct" justice into ethics.[26] However, it does not look as if he can say anything substantial about how to move in the other direction—how to generate justice from ethics.

There are two schools of thought on this issue. Levinas's critics fault him for his inability to offer positive claims about what norms express the responsibility that emerges in the encounter with the other. His defenders also accept that he cannot specify any sort of normative ethics. They simply do not consider this to be a flaw. Levinas's great insight is that our social institutions rest on a feature of human existence that they cannot help but obscure. His silence in not offering an account of justice simply shows that he remains faithful to the great insight of his philosophy.

Many also take his reticence to be expressive of a broader stylistic device. Levinas's prose is notoriously obscure. Many allege that his writing style is a consequence of his claims about the transcendence of the human other. Whiffs of this view can be found in Jill Robbins's *Altered Reading*. Robbins claims, reiterating proposals by Jean François Lyotard and Jacques Derrida, that the last thing we should do as interpreters of Levinas is subject him to straightforward exposition. "The best way to respect his work," she notes, "is to read it deviously, to misunderstand it."[27] Her suggestion is motivated partly by the observation that Levinas does not present his ideas through the straightforward, linear arguments one often finds in philosophic texts. His writing is more poetic, more evocative, what he might call "prophetic." Her further thought, presumably, is that he adopts this style because he views himself as writing about matters that cannot be put into words. It is his way of highlighting the inadequacy of language, at least the language of philosophy, to do justice to the human other's ineffability.

Levinas's reticence with respect to justice can be interpreted as part of this stylistic device. We may not be able to do without justice or moral norms, but they also cannot help but misrepresent the primordial ethical experience they are supposed to reflect and protect. Levinas's aim in not advancing an account of justice is to stress this idea that our normative ethical theories, our notions of justice, are at best necessary evils that we must continually subject to critical scrutiny, and if he were to advance one of his own, it would reinforce the misguided belief that it is possible to develop a nonmisleading theory of justice. He would be inviting us, in effect, to commit what he considers to be the classic crime of philosophers: the mistake of trusting in our ability to comprehend all things and, in doing so, neglecting the ineffability of the other person. By not offering an account of justice, he draws attention to the fact that normative ethical theories rest on a foundation they cannot help but misrepresent.

So, there seem to be two schools of thought on Levinas's silence. His defenders argue that his silence is indicative of his sensitivity to the limits of philosophy with respect to the ethical. It is similar to Wittgenstein's silence in the *Tractatus Logico-Philosophicus* when Wittgenstein quips in his preface that the greatest insight of the *Tractatus* rests in what it does not say, in how it respects the ineffability of ethics by not speculating about ethical questions.[28] So, too, Levinas's failure to *assert* a normative ethics is his way of *showing* the ineffability of the ethical. The other school of thought questions this quietism. Silence may be a fitting tribute to the ethical, but the world is drenched in injustices so galling it is irresponsible *not* to speak out against them. Here it is useful to juxtapose Wittgenstein's silence with a passage from Augustine's *Confessions* that Wittgenstein admired. "Can anyone speak enough of you?" Augustine asks God at the start of the *Confessions*, "Yet woe betide those who are silent about you!"[29] Augustine's point is that nothing he says about God can do justice to God's magnificence. Yet it is not available to him to remain quiet; it would be *irresponsible* not to speak out in praise of God. Similarly, we should expect Levinas to say more about what normative ethics he favors, critics like Critchley allege, even if doing so requires him to violate his thesis concerning the human other's ineffability.[30]

6. Why Is It Hard to Talk about Justice?

More could be said about the nuances of each of these stances toward Levinas. I want to resist the temptation, though, to become embroiled in a debate over their intricacies. Instead, I want to focus on what both take to be uncontroversial claims to make about Levinas. Both accept what I earlier called the "ineffability hypothesis." The ineffability hypothesis says that Levinas does not advance a normative ethics because to do so would require him to violate his core claims about the transcendence of the other. Levinas's defenders and critics alike both accept this claim. They simply disagree about the acceptability of this silence.

My suggestion in this book is that we do not need to accept the ineffability hypothesis. Nothing in Levinas's philosophy prohibits him from generating ethical norms that are based on the encounter with the other. We can talk about our experience of the human other in these paradigmatic moral moments in clear, unambiguous terms. *And* we can do so without misrepresenting it. Someone familiar with Levinas will no doubt find these claims controversial. His thesis, after all, is that "the human other is absolutely other," and he repeatedly cautions philosophers against presuming to comprehend the human other. He reiterates this warning time and time again, stressing that ordinary language and the technical language of philosophy cannot do justice to the other's transcendence. I seem to be claiming that we can comprehend and describe the other, whereas the whole point of his philosophy is to discourage this assumption.

I agree with this view up to a point: Levinas's main thesis is that the human other is absolutely other. However, I disagree about what is alleged by this thesis and why it is provocative. Levinas's admirers often seem to be drawn to him because of the passion with which he attacks philosophers for daring to theorize about human existence; his bold idea is that there is a dimension of human life that can never be expressed in language or social institutions. However, this idea is not novel. Several of Levinas's contemporaries, such as Bataille and Blanchot, wrote about limit experiences that defy comprehension. The idea that human existence cannot be fully explained by any philosophy is also a staple idea of existentialism. So, if Levinas is novel, it is not because of the fervor of his critique of philosophy. Plenty of philosophers before and after him have been equally vexed by their peers' hubris in ignoring the fact that there is more between heaven and earth than they can dream in their work. If Levinas is novel, it is because he cites *ethically momentous events* as being, in some sense, transcendent.

So, I find myself inclined to interpret his thesis that the "human other" is "absolutely other" differently than do others. *Autre* is usually read as the key term in his thesis that *l'absolument Autre, c'est Autrui*. However, my sense is that *Autrui* is doing the interesting explanatory work. Levinas is concerned to stress not the utter transcendence of the human but, as it were, the utter humanity of transcendence. *Autrui* is the *explanans*, and *Autre* is the *explanandum*.

Let me amplify on this alternative approach. I noted before that *autrui* corresponds to our use of *others* to refer to *other persons*. Levinas's peers also used it in discussing the problem of other minds. Levinas expressly distinguishes his use of this term, however, from its usage in these debates. He insists that "the human other" (*autrui*) is not an "alter ego," literally "another I," another mind or consciousness, and he stresses instead that "[t]he Other is the weak, the poor, the widow and the orphan, whereas I am the rich or the powerful."[31] He also notes that "to recognize the other is to recognize a hunger" and that "to recognize the other is to give."[32] So, Levinas uses *autrui*, in keeping with its standard usage, to refer to other persons, but he also offers a suggestion as to what situations most perspicuously reveal what it means to be a person. His interesting claim is that they are ethical in character.

It helps to compare Levinas to Kant on this point. Kant discusses the feeling of respect (*Achtung*) in the *Groundwork* and *Critique of Practical Reason*. He believes that the respect we feel for others gives us an insight into our nature; properly scrutinized, it reveals that we are fundamentally autonomous beings. Levinas can be read as making a similar point, albeit with something like pity or compassion taking the place of Kantian respect. Levinas thinks that philosophers have erred in focusing on our cognitive abilities in theorizing about personhood. In particular, they too often conclude that the best way to determine what is essential to humanity is to reflect on one's experience of one's own nature. Thus I directly know, apropos of Descartes, that I am a thinking thing, and I conclude on the basis of this insight that talk about humanity or personhood boils down to talk about other minds. Levinas upholds a different strategy, one that is more in keeping with his training in phenomenology. He begins with this question: when are we most vividly struck by another person's presence *as another person*? He follows Kant in suggesting that such moments are ethical. I am most powerfully awed by another person's presence when the sight of someone humiliated fills me with shame. Or it comes in times of great crisis—such as, for Levinas, the Holocaust—when I feel a sense of outrage at the victimization of others. It is in these moments that we are led to appreciate what is precious and unique about human beings.

So, Levinas uses *autrui* to refer to our recognition of another person's personhood, but with the caveat that humanity is most fully revealed in moments of compassion. He draws the corresponding conclusion that humanity is, fundamentally, something that inspires respect. To recognize the existence of another person is not to recognize another mind but a source of responsibility, something one finds oneself impelled to treat decently. This analysis of what Levinas means by *autrui* changes how I read his slogan that "the human other is absolutely other." His point in declaring the "human other" to be "absolutely other" is to emphasize how acknowledging humanity inevitably has a practical side to it. Again, another person is not so much something one *knows* as something one finds oneself obligated to *help*, and the true test of whether one has grasped this fact is not borne out in claims to *understand* it but in the *deeds* one performs in response to it, in the life one lives, in whether one does, in fact, help others. Levinas insists on the other's incomprehensibility, then, because he finds it misleading to depict interpersonal relationships as primarily involving comprehension in the first place.

I defend a different reading of Levinas, then, from the one I find reflected in the ineffability hypothesis. Levinas is often read as if his epistemic claims about the other's incomprehensibility were the foundational idea of his philosophy, and his remarks on ethics are depicted as consequences of them. However, his argument moves in the other direction: he begins with the idea that we cannot help but recognize one another as unconditionally valuable, and he draws his epistemic conclusions about interpersonal relations surpassing comprehension to stress how acknowledging this dignity differs from other acts of cognition—how it requires a practical response.

My goal in this book, at any rate, is to convince you that Levinas should be read in this way. I would note, though, in making an initial plea for it that it lets me address a basic question about his philosophy. Someone unsympathetic to Levinas might ask the following: Why should we think there is anything transcendent or "absolutely other" about other persons in the first place? This is a bold claim. What can Levinas say to convince us that it is true? I am not sure if the standard approaches to him, approaches that take his epistemic claims about the other's incomprehensibility as his foundational idea, have an answer to this question.

My sense is that they find it either uncontroversial or so foundational to his thinking that it cannot be defended. We simply need to grant Levinas this point. Once we do, he has interesting things to say about how philosophers have failed to respect the ineffability of the other person. On my reading, however, he has an argument for this claim. He has a story to tell about why we cannot help but acknowledge others as sources of responsibility, and he has an interesting story to tell about the constraints this acknowledgement imposes on us, how it requires us to view them as being in some sense miraculous.

My goal is to share this story. The point I want to stress for now is simply that we cannot assess the accuracy of the criticisms leveled against Levinas until we hear it. Levinas's critics fault him for not advancing anything like a normative ethics. Yet we saw that this criticism is misleading: their complaint is not that he *does not* but that he *cannot* offer any such ethics. We also saw that this criticism relies on the ineffability hypothesis, which alleges that he cannot advance any substantive, action-guiding ethical principles, procedures, or norms without violating his claims about the other's transcendence. However, we cannot gauge the accuracy of this hypothesis until we know more about his rationale for deeming the human other to be transcendent in the first place. Put simply, we cannot say what follows from his thesis that *l'absolument Autre, c'est Autrui* until we know what is being alleged in it. My claim in this book is that Levinas reaches this conclusion by reflecting on the challenges involved in acknowledging others as ethically invaluable. My further claim is that when we clarify these challenges, we will see that they in no way preclude Levinas from developing a normative ethics.

I will turn to establishing these two claims in the next chapter, where I provide an analysis of a pivotal section of Levinas's *Totality and Infinity*, his "murder argument" in the "Ethics and the Face" chapter of section 3. Before I turn to this reading of *Totality and Infinity*, however, let me consider an objection against my argument in this chapter. My key claim in chapter 1 is that Levinas's critics and defenders alike both accept the ineffability hypothesis; both agree that Levinas cannot specify a normative ethics or an account of justice without violating his claims about the human other's ineffability. One could argue, however, that I am creating a straw man. I claim that Levinas's readers accept that the human other is ineffable without supplying examples from the secondary literature to support this allegation, and this omission allows me to illicitly frame the debate on my own terms.

This worry is legitimate. I find it difficult, however, to offer concise examples from the secondary literature where commentators expressly argue for the ineffability hypothesis. Earlier, I quoted essays by Richard Rorty and by Simon Critchley, where they fault Levinas for defending a conception of ethics that cannot be put into practice, and I suggested that their commitment to the ineffability hypothesis is what drives them to voice this complaint. Alain Badiou faults Levinas for similar reasons, I would argue, in *Ethics: An Essay on the Understanding of Evil*. Elsewhere in this book, I consider texts by Axel Honneth and Jill Robbins where, I find, they assume something like the ineffability hypothesis in their discussions on Levinas.

Nonetheless, it is difficult to offer examples of the ineffability hypothesis because what I have in mind by it is not so much an acknowledged school of thought about how to read Levinas but a cluster of widely accepted assumptions about his philosophy in general. It seems to me that most commentators accept his claim that the human other is "absolutely other" as uncontroversial and as preinterpretative. This is not to say that they do not discuss the subtle ways in which the face of the other resists comprehension. It is not uncommon, for example, to find probing analyses of the inadequacy of Husserlian phenomenology to do justice to the relation to the other, or of how various forms of language or experiences of time misrepresent it. Nonetheless, most commentators do not question Levinas's central thesis that the other is transcendent; they focus instead on elaborating the consequences of it. Moreover, it seems to me that many write about the relation to the other as though it were ineffable, as it were, *simpliciter*—a relation that stands beyond all possible comprehension or expression, without distinguishing different ways or extents to which it might resist comprehension or expression.

Consider, along these lines, the following passages from Bernhard Waldenfels's "Levinas and the Face of the Other" and Edith Wyschogrod's "Language and Alterity in the Thought of Levinas":

> If the other's face transcends the ontological reign of more or less defined entities we are able only to say what it is not, or more precisely: we can only show that it is not something at all. The list of negation is long and sometimes tiresome. We are told that the face not something we can see or touch, while moving within open horizons, passing through changing perspectives, transforming it into a content we embrace or manipulate...It has no 'plastic' form to be transformed in images; it has no *eidos*, no 'adequate idea' by which we could represent or grasp it.[33]

> The face is not an appearance but rather an epiphany that resists conceptual grasp, rending the sensible through which it appears. It proffers itself as defenceless, "in the nudity of the absolute openness of transcendent" (TI 199) expressing in its alterity as destitution and as a solicitation to desist from violence.[34]

Waldenfels's and Wyschogrod's essays are taken from *The Cambridge Companion to Levinas*, and I select them because an aim of the Cambridge Companion series is to publish reference works for nonspecialists. These essays, then, are texts by recognized scholars which aim at, among other things, explicating key areas of Levinas's thinking to nonspecialists. Waldenfels and Wyschogrod are concerned in the above quotations to explain Levinas's idea of the face (*le visage*) of the other, particularly the idea that the face resists comprehension. I take Waldenfels's characterization, in the first, to be emblematic of the ineffability hypothesis: the other's face is radically inexpressible and incomprehensible, a "not something at all." Wyschogrod does not say outright that other is ineffable, but I find it telling that she glosses Levinas's reasons for alleging that the face cannot be comprehended—that it is an "enigma" not an "appearance"—by using Levinas's language of defenselessness and destitution without explaining how it supports the idea that the other "resists conceptual grasp."

Now, I do not want to overstate this criticism. Waldenfels and Wyschogrod are both concerned to explain why the other resists comprehension, with Waldenfels focusing on the idea of the face and with Wyschogrod discussing this issue as it relates more generally to language. Nonetheless, neither is concerned to write a primer on Levinas on this issue, and my aim is not to fault their essays. I quote them merely to give examples of how the notion that the other is ineffable filters through the secondary scholarship as a preinterpretative assumption. Wyschogrod's use of Levinas's language in what purports to be an explanation *of this language* strikes me as revealing; his idea that the other resists conceptualization because of her destitution is not so much *explained* as *reiterated*. My sense is that this type of move, in which Levinas's claims are restated in what purport to be explanations of them, is fairly common in exegetical pieces, and one effect is to engender the sense that the other is ineffable *simpliciter*. *Why* the relation to the other resists conceptualization goes unexplained, encouraging us to think that it is preinterpretative, that it must be radically ineffable. Moreover, when commentators expressly take up this question, as Waldenfel does in the first quotation, they characterize the relation to the other in terms of ineffability, to depict the other as radically indescribable, incomprehensible, inexpressible, and so forth.

There is, however, at least one notable exception to this tendency, one I associate with more deconstructionist readings of Levinas.[35] One could argue that the relation to the other is not *ineffable* but *overdeterminate*. That is, the relation to the other resists conceptual grasp not because it stands beyond all comprehension, description, or expression, hence beyond all codification in ethical norms, but because it generates too many conflicting ethical norms. Levinas could here be read as defending the ethical equivalent of deconstructionist literary criticism. Just as the deconstructionist seeks to show how evidence can always be culled from a text that supports multiple and conflicting interpretations, so, too, it could be argued that the human other is "absolutely other" in the sense that multiple, conflicting moral norms can be derived from the encounter with the other.

This is a good objection, one I take up in the final chapter of this book, where I conclude by arguing for a broadly pragmatic interpretation of

Levinas. I want to assure the reader, therefore, that my intent is not to make a straw man out of my opponents. For now, I find it useful to write as if Levinas's critics and defenders are committed to the idea that the human other is ineffable, and they establish their respective criticisms and defenses of him based on their belief that he cannot give a positive account of justice because the other is ineffable. My own personal sense is that something like the ineffability hypothesis is accepted by many of Levinas's readers—that it designates a cluster of preinterpretative assumptions that are widely accepted in the secondary literature and that serve as a basis upon which commentators draw conclusions about other areas of his thinking. However, it would be obtuse not to admit that there are more subtle analyses of why the other resists conceptualization, and I will address them as my own position becomes clearer—first, in chapter 3, by distinguishing it from that of Jean François Lyotard in *Just Gaming*,[36] and later, in chapter 4, by defending my more pragmatic approach to Levinas on this issue against a deconstructionist approach that associates the relation to the other with overdetermination.

Endnotes

1. Axel Honneth characterizes Levinas's ethics as "an ethics of unlimited care" in "The Other of Justice" in *Cambridge Companion to Habermas*, ed. Stephen K. White (Cambridge: Cambridge University Press, 1995).
2. Richard Rorty, *Achieving Our Country: Leftist Thought in Twentieth-Century America* (Cambridge, MA: Harvard University Press, 1998), 96–97. Alain Badiou also faults Levinas for having a uselessly impractical ethical theory in *Ethics: An Essay on the Understanding of Evil*, trans. Peter Hallward (New York: Verso, 2001).
3. Critchley argues in several places that Levinas has interesting things to say about how social and political institutions inevitably misrepresent the ethical relations they are designed to protect, and thus it is possible to develop a certain form of political critique based on his work. Critchley also thinks, though, that we are entitled to expect more from Levinas. That is, we need not only tools for criticizing existing political institutions, but also advice on how to generate more humane ones in their place. He faults Levinas for not offering more advice on this score.
4. See Critchley, "Habermas and Derrida Get Married," in *The Ethics of Deconstruction* (Edinburgh: Edinburgh University Press, 1999), 267–280. *See esp.* p. 269.
5. See Critchley's introduction to *The Cambridge Companion to Levinas*, ed. Simon Critchley and Robert Bernasconi (Cambridge: Cambridge University Press, 2002), 1–32. *See esp.* p. 27.
6. Critchley, "Habermas and Derrida Get Married," 275.
7. Critchley, "Five Problems in Levinas's View of Politics," 173.
8. See, however, Critchley's "Five Problems in Levinas's View of Politics," 172–173. Some of his complaints about the current state of Levinas scholarship, particularly its scholasticism, suggest that he may not find this proposal so troubling after all.
9. Critchley, *Cambridge Companion to Levinas*, 27.
10. See Critchley, *Cambridge Companion to Levinas*, 27–28. Critchley borrows this distinction from Hilary Putnam, who, in turn, borrows it from Stanley Cavell. Putnam characterizes Levinas as a "moral perfectionist" in "Levinas and Judaism," 36–37. Cavell originally coined this distinction in *Conditions Handsome and Unhandsome* (Chicago: University of Chicago Press, 1990).

11. Kant, *Groundwork of the Metaphysics of Morals*, trans. H. J. Patton (New York: Harper Torchbooks, 1964), 55.
12. Robert Bernasconi proposes something like this interpretation of Levinas in Bernasconi, "What is the question to which 'substitution' is the answer?" *Cambridge Companion to Levinas*, ed. Simon Critchley and Robert Bernasconi (Cambridge: Cambridge University Press, 2002), 234–251. Bernasconi here argues that Levinas's aim in his later writings on "substitution" is to "address question of what the subject must be like for ethics to be possible."
13. See, for example, Levinas's discussion of "war" in the Preface of *Totality and Infinity*, 21–23. See also the following texts for dismissive references to Hobbes: *Otherwise than Being or Beyond Essence*, 159; "The Rights of the Other Man," in *Alterity and Transcendence*, trans. Michael B. Smith (New York: Columbia University Press, 1999), 145–149. *See esp.* p. 147.
14. See, for example, Levinas, "Peace and Proximity," in *Basic Philosophical Writings*, ed. Adriaan T. Peperzak, Simon Critchley, and Robert Bernasconi (Bloomington, IN: Indiana University Press, 1996), 161–169. *See esp.* pp. 164–165.
15. Here I would add that I think that the same should be said of moments when Levinas makes claims about the political events of his time that strike contemporary readers as dubious or uninformed. Simon Critchley and Howard Caygill, for example, have drawn attention to places in Levinas's oeuvre—particularly interviews and nonphilosophic writings—where, they argue, he defends naïve theses concerning the state of Israel and the Israeli-Palestinian conflict. Both suggest that his failure to do justice to the complexity of these issues is symptomatic of larger failings in his conception of justice and politics. My own view is that his remarks on the Israeli-Palestinian conflict are more balanced than Critchley and Caygill admit, but I also question whether they should be read as symptomatic of larger flaws in his thinking.

    For example, Levinas once remarked in an interview that non-Western conceptions of "humanity" were mere "dance." This claim strikes me as ethnocentric, and it is disappointing that he said it. However, I do not see that his personal ethnocentrism *itself* constitutes grounds for rejecting his analysis of ethics. A comparison with Wittgenstein may prove helpful. I find it disappointing that Wittgenstein says so little about ethics in his *Philosophical Investigation*, given the fecundity of his ideas in these notes. He also defended some wildly uninformed ideas about politics in his lifetime.

However, I do not see why these disappointments should, of themselves, lead me to reject his claims about language in the *Philosophic Investigations*. Similarly, I do not see why the absence of a full-blown theory of justice in Levinas, or the fact that he occasionally held questionable beliefs about the political events of his time, should lead me to reject his entire account of ethics. The appropriate response is to feel disappointed. To be clear, my intent is not to romanticize Levinas or to lapse into uncritical discipleship. There are grounds for feeling disappointed by various areas of his thinking. I simply question efforts to mine this disappointment for broader flaws that are supposedly endemic to his philosophy.

See Mortley, *French Philosophers in Conversation* (London: Routledge, 1991), 18, for Levinas's characterization of non-Western views of humanity as "dance." (To his credit, Levinas cautions that it may be offensive, or "dangerous," to "publicly" state this belief.) See Critchley, "Five Problems in Levinas's View of Politics," 175, 176, for, respectively, a critique of Levinas's view of Israel and a critique of his ethnocentrism. See Caygill, *Levinas and the Political* (London: Routledge, 2002), 159–198, for a critical analysis of Levinas's views concerning Israel.

16. Levinas indicates his debt to Kierkegaard in "Existence and Ethics," in *Proper Names*, trans. Michael Smith (Stanford: Stanford University Press, 1996), 26–38, where he presents himself as modifying Kierkegaard's critique of Hegel. He also cites Kierkegaard as an influence in "Enigma and Phenomena," in *Basic Philosophic Writings*, ed. Adriaan T. Peperzak, Simon Critchley, and Robert Bernasconi (Bloomington, IN: Indiana University Press, 1996), 65–77. *See esp.* p. 76. He remarks on his debt to Rosenzweig in the opening pages of *Totality and Infinity*, explaining that Rosenzweig's *Stern der Erlosung* is "too often present in this book to be cited." See Levinas, *Totality and Infinity*, 28.
17. This claim dates back to Parmenides. Levinas often cites Parmenides as the founding father of the philosophic tradition he is chiefly concerned to criticize. See the following for examples of Levinas's worries about Parmenides's/Hegel's dictum: *Time and the Other*, 8, 42, 43; "God and Philosophy," in *Of God Who Comes to Mind*, trans. Bettina Bergo (Stanford: Stanford University Press, 1998), 55–78. *See esp.* pp. 77–78.
18. An example of this type of argument can be found in Levinas's "Transcendence and Height" (1962), which is generally considered to be one of the best synopses of *Totality and Infinity*. See "Transcendence and Height," 13–14.

19. I am indebted to Simon Critchley for clarifying the source of these two terms in Plato. See his introductory remarks in Levinas, *Basic Philosophical Writings*, ed. Adriaan T. Peperzak, Simon Critchley, and Robert Bernasconi (Bloomington, IN: Indiana University Press, 1996), x–xi.
20. For example, see Levinas, *Time and the Other*, 43, 75.
21. See Levinas, "Transcendence and Height," 14.
22. Ibid.
23. See Levinas, *Totality and Infinity*, 39.
24. I am being deliberately terse in explaining this last idea in Levinas—his thesis that "the human other is absolutely other." I want to avoid explaining it in detail for the time being for two reasons. First, my aim is to give only a brief overview of Levinas's philosophy. I need to help you understand just enough of his thinking so that his critics' complaints will make sense. Suffice to say, then, that I will explain his thesis that "the human other is absolutely other" in greater detail in the next chapter. Second, my sense is that this claim is both the most important and the least well-understood idea in Levinas's philosophy. My chief aim in chapters 3 and 4 will be to present an interpretation of why Levinas associates humanity with transcendence. This topic, then, is extremely complex, and I want to postpone discussing it in detail until later in this book.
25. Critchley, *Cambridge Companion*, 27.
26. Critchley argues in several places that Levinas can be read as developing this kind of "ethical deconstruction," most notably in his book *The Ethics of Deconstruction*.
27. Jill Robbins, *Altered Reading* (Chicago: University of Chicago Press, 1999), 13. Robbins is here summarizing what she takes to be a fairly standard way of reading Levinas among recent French philosophers. She frequently cites, for example, Jacques Derrida and Jean-François Lyotard in setting out her interpretation of him.
28. See Wittgenstein *Tractatus Logico-Philosophicus*, trans. D. F. Pears and B. F. McGuinness (London: Routledge, 1961), 5. See Plant's *Wittgenstein and Levinas: Ethical and Religious Thought* (London: Routledge, 2005) for more discussion of possible similarities between Levinas and Wittgenstein.
29. Augustine, *Confessions*, trans. R. S. Pine-Coffin (New York: Penguin Books, 1961), 23.
30. I should note, in fairness to Critchley, that his position on this issue is far more complex than this sketch suggests. It may help to say a few words by way of introduction to his interpretation of Levinas, especially given how large a role it plays in this book as a target of criticism.

Critchley was one of the first commentators to argue for an "ethical turn" in deconstructionism. That is, he was among the first to question the widespread view of deconstructionism as an apolitical, nihilistic philosophy. Critchley demonstrated, most notably in *The Ethics of Deconstruction*, that deconstructionist critical strategies are grounded in an understanding of ethics, one Derrida inherits from Levinas, and that they are compatible with forms of political critique. Deconstructionists view systems of justice as rooted in an ethical relationship to the other that can never be fully recuperated in their norms, standards, and institutions. The deconstructionist attempts to show, in turn, how existing theories of justice invariably misrepresent the ethical relationship to the other that inspires them. Critchley has defended this understanding of deconstructionism in several writings, and thus he has frequently defended Levinas against critics who claim that his work has no practical value whatsoever. Elsewhere, however, he has raised the question of whether Levinas's understanding of justice is adequate and whether it needs to be supplemented with a more procedural normative ethical theory.

At times, then, Critchley can sound like one of Levinas's admirers, praising him for resisting the misguided urge to advance a positive account of justice. At other times, he can sound like one of his critics, faulting him on this same point. He criticizes Levinas for having an inadequate account of justice, in his Introduction to *The Cambridge Companion to Levinas*, in his essay "Habermas and Derrida Get Married," and in "Five Problems in Levinas's View of Politics." Elsewhere, however, he defends Levinas against Rorty's accusation that Levinas's philosophy is a "nuisance" with no practical value whatsoever. See, for example, his exchanges with Richard Rorty in *Deconstruction and Pragmatism*, ed. Chantal Mouffe (London, Routledge, 1996).

In his most recent work, *Infinitely Demanding: Ethics of Commitment, Politics of Resistance* (New York: Verso, 2007) and in the essay "Five Problems in Levinas's View of Politics," Critchley defends a form of anarchism as ethical practice that seems to be similar to the supportive view of Levinas that I discuss in this section.

31. See Levinas, *Time and the Other*, 83.
32. Levinas, *Totality and Infinity*, 75.
33. Waldenfels, "Levinas and the Face of the Other," in *Cambridge Companion to Levinas*, ed. Simon Critchley and Robert Bernasconi (Cambridge: Cambridge Unviersity Press, 2002), 63–81. *See esp.* p. 66.
34. Wyschogrod, "Language and Alterity in the Thought of Levinas," in *Cambridge Companion to Levinas*, ed. Simon Critchley and Robert Bernasconi (Cambridge: Cambridge University Press, 2002), 188–205. *See esp.* 195–196.

35. My sense is that Robert Bernaconi's "Rereading Levinas," in *The Question of the Other: Essays in Contemporary Continental Philosophy* (Albany: SUNY Press, 1989), is the unacknowledged *locus classicus* for this interpretation of Levinas in the English-language scholarship on him. I discuss Bernasconi's essay in the final chapter of this book, where I argue against his deconstructionist reading of Levinas and for a more pragmatic interpretation. I argued earlier in this chapter that Simon Critchley's criticisms of Levinas draw on the ineffability hypothesis. My sense is that this criticism may hold true for those essays where he faults Levinas for not offering a more robust account of justice, for in them he does not amplify on why Levinas's characterization of the relation to the other prohibits him from developing a theory of justice. However, it seems to me that he defends a deconstructionist reading of Levinas, one that associates the relation to the other with overdetermination rather than ineffability, in *The Ethics of Deconstruction*.
36. Lyotard's remarks in *Just Gaming* strike me as constituting a third stance on why the other cannot be comprehended. My sense is that much of the literature takes this claim to be preinterpretative and assumes that the other is ineffable. A deconstructionist reading would distinguish ineffability from overdetermination. By contrast, Lyotard associates the encounter with the other with a difference between prescriptive and descriptive language games. He suggests that the encounter with the other is inherently prescriptive—that it amounts to a lived "prescription that there be prescriptions." The other resists conceptualization, in turn, because it is impossible to capture this prescriptive command using descriptive language. My own reading of Levinas, we will see, is similar in that I too read Levinas as stressing the distinction between prescription and description, ethics and theory, agency and spectatorship, when he insists on the human other's "absolute otherness." However, I am less pessimistic than Lyotard about the possibility of generating a theory of justice based on the ethical encounter with the other.

Chapter 2

Murder and Moral Realism

Everything, even his crime, his sentence and imprisonment, seemed to him now in the first rush of feeling an external, strange fact with which he had no concern...Life had stepped into the place of theory and something quite different would work itself out in his mind.

—Dostoevsky, *Crime and Punishment*

1. Introduction

I argued in the last chapter that the criticism that Levinas's remarks on ethics have little practical value is apt to look unavoidable if we accept what I call "the ineffability hypothesis." Levinas insists that other persons "overflow comprehension"[1] and embody what he variously calls infinity, mystery, enigma, transcendence, the uncontainable, the ungraspable, the unforeseen, and the absolutely other. His critics and defenders alike conclude that he cannot specify any norms or procedures for determining ethical conduct. To do so, he would need to subject interpersonal relations to some measure of description or exposition, which, by hypothesis, he cannot do since they bring us into relation with something ineffable.

I closed chapter 1 by suggesting that we need to ask why Levinas identifies humanity with infinitude or transcendence in the first place. In particular, we need to ask what role ethical responsibility plays in grounding this association. Levinas writes in "Transcendence and Height" (1962) that "the infinite consists in the impossibility of escaping from responsibility,"[2] and he observes in *Totality and Infinity* that the "strangeness of the other, his irreducibility to the I, to my thoughts and my possessions, is precisely accomplished as a calling into question of my spontaneity, as ethics."[3] Levinas thinks that there is something about accepting responsibility for others that requires us to view them as miraculous or infinite. Ethics entails transcendence. My task in this chapter is to explain this connection.

This is not an easy task. Levinas gives clues, however, as to the link between ethics and transcendence in two important texts. His 1951 essay "Is Ontology Fundamental?" is the first place where he remarks on the importance of ethical responsibility, and he associates it with the claim that the "human other" (*autrui*) is "absolutely other" (*autre*). He discusses this claim in significantly greater detail in his first major work, *Totality and Infinity* (1961). In both texts, Levinas also remarks on "the ethical impossibility of murder." It is impossible, he argues, to murder another person: there is something about acknowledging another person *as another person* that requires us to regard her as inviolable. My suggestion is that Levinas intends for this argument to establish his thesis about the transcendence of the human other. If we want to understand the link between ethics and transcendence, we should look to his murder argument.

Before I delve into the details of Levinas's argument, let me say more about why it merits attention. I will be putting *a lot* of interpretive weight on this argument. Yet it appears somewhat infrequently in his oeuvre. His writings are peppered with references to murder and violence.[4] However, he only expressly argues that murder is impossible, to my knowledge, in four texts from the middle period of his career: "Is Ontology Fundamental?" (1951), "Ethics and Spirit" (1952), "Freedom and Command" (1953), and *Totality and Infinity* (1961). So, it may seem that I am making too much of this argument, treating a passing idea as if it held the key to Levinas's entire philosophy. There are indications, though, that Levinas's murder argument plays a particularly important role in his mature philosophy.

Levinas's writings from the fifties and early sixties occupy a special place in his oeuvre. It is with them that he established himself as an independent thinker, not merely an exponent and popularizer of Husserlian phenomenology. "Is Ontology Fundamental?" and *Totality and Infinity* occupy a special place, in turn, among the writings of this period. Levinas had flirted prior to them with the idea that our lives are shaped by a certain longing for transcendence, for what he terms, as early as 1935, a "new path out of being."[5] He had also begun to worry about the failure of philosophers to make sense of this longing. He had even argued in a series of talks in 1947, published as *Time and the Other*, that it is primarily in our interactions with other persons that we brush up against something absolutely other. What distinguishes his work in the fifties and early sixties is his suggestion that it is through our *responsibility* for others that, as he puts it in *Totality and Infinity*, such infinitude is "concretized" or "deformalized."[6] "Is Ontology Fundamental?" is the first place where he advances this claim. *Totality and Infinity* is his most complete expression of it. It is revealing that he argues for the impossibility of murder in each of these crucial texts.

The murder argument also occurs at pivotal moments in each text. "Is Ontology Fundamental?" contains five short sections. Levinas assesses Heidegger's philosophy in the first three sections of his essay. It is not until the fourth section that he introduces his thesis that "the human other" is "absolutely other," and he does not defend this claim until the final section of his essay, section 5, where he associates it with the "ethical signification of the other." He offers his murder argument to explain "ethical signification." The murder argument is positioned, then, as his justification for the claim that human other (*autrui*) is absolutely other (*l'absolument Autre*).

Totality and Infinity is a larger and more detailed work, but it has a similar structure. The book consists of four main sections. Levinas gives an overview in section 1, "The Same and the Other," of his book's main problem and his proposed solution. He is interested in clarifying how it is possible to interact with something genuinely transcendent—how, as he puts it, we can enter into a "relation with a reality infinitely distant from my own reality, yet without this distance destroying this relation and without this relation destroying this distance."[7] He also hopes to show that such a relation is effected through responsibility for others. He does not defend this last claim, however, until section 3, "Exteriority and the Face." Section 2, "Interiority and Economy," is devoted to an analysis of what he calls "enjoyment" (*jouissance*)—our everyday, preethical mode of existence, the sense we have of our place in the world that is interrupted when we recognize ourselves as responsible for another's well-being. When Levinas returns to his suggestion that the human other is absolutely other in section 3, he defends this claim in a chapter titled "Ethics and the Face" by reiterating his murder argument. His murder argument is presented as his explanation for why accepting responsibility for others requires us to view them as miraculous, then, in what are arguably the two most important texts of the most important period in his philosophy. One can look elsewhere in his corpus for suggestions as to why ethics and transcendence might be linked, but his murder argument stands out as a particularly apt starting point for thinking about this connection.

Let me say more about the role of Levinas's murder argument in the structure of *Totality and Infinity*. Levinas's aim in *Totality and Infinity*, again, is to show how it is possible to interact with "a reality infinitely distant from my own reality, yet without this distance destroying this relation and without this relation destroying this distance." For all his fervent remarks on pity, then, the explicit aim of *Totality and Infinity* is fairly abstract: Levinas wants to prove the existence of something like mind-independent reality. He worries that philosophy has tended to lapse into a kind of idealism, not one that denies the existence of mind-independent reality as such, but one that reduces reality to a series of rational processes, to a manifestation of Mind or what he calls "the Same." The goal of *Totality and Infinity*, in turn, is to determine if we experience things in our lives that cannot be so reduced. Thus Levinas subtitles his book "an essay on exteriority." His aim is to document the presence in our lives of at least one thing that remains "exterior" to any theory we construct about it.

If this gloss is correct, then Levinas needs to do two things in his book. He needs to show that we experience things that cannot be fully comprehended; he needs to show, that is, that we experience things for which it is implausible to think we could have authored them ourselves using any philosophic system. This first task is fairly difficult since it requires him to show that there is something such that (1) we are conscious of its presence without (2) being able to cognize it fully. However, Levinas also needs to prove, second, that an entity actually exists that instills this response in us. After all, the mere fact that I cannot conceive something does not imply that it exists. I cannot conceive how it can both rain and not rain at the same time, but the mere fact that I cannot conceive *p and ~p* does not imply that somewhere it is both raining and not raining. So, too, it is not enough for Levinas merely to document moments when we *seem* to brush up against something transcendent; he needs to establish that something transcendent is, so to speak, really out there doing the work, prompting this response from us.

I stress this point so as to reiterate my earlier observation about the structure of *Totality and Infinity*. Levinas opens *Totality and Infinity* by giving an overview of his book's main problem; he depicts it as an attempt to establish the existence of something like mind-independent reality, "exteriority." He does not share his solution, however, until the third section of *Totality and Infinity*, where, as I noted, he gives his murder argument. However, there is a further point to note here as well. Commentators on Levinas have not appreciated the rigor, I think, of his solution to his book's opening problem. Many have noted that he thinks we experience something that cannot be fully comprehended in our interactions with other persons; many have noted that he titles his book "an essay on exteriority." However, none has observed, to my knowledge, that he needs to demonstrate the actual *existence* of something transcendent to fulfill his book's promise. They have been so smitten, it seems, by the force of his claims on behalf of the *incomprehensibility* of interpersonal relations that they have overlooked the fact that he must also establish the *reality* of such relations. We will see, though, that one of the chief aims of his murder argument is precisely to establish this point.

Levinas focuses on murder because it strikes him as the most flagrant example of a case where we seem to be able to disregard the dignity of human life. Someone who commits murder tries, after all, to destroy another person's life. So, our ability to perform this act seems to suggest that we do not necessarily recognize one another as possessing any sort of inalienable dignity. Levinas thinks he can show, though, that an appreciation of others as possessing such dignity is presupposed even in this most extreme case. Murder is impossible in the sense that the drive to violate others is invariably haunted by an awareness of them as incontestably precious. Notice, though, that this awareness of others should be inescapable if Levinas can establish this point. For if even the most determined efforts to ignore it fail, then it seems that we cannot help but tacitly acknowledge it in less brazen cases. Levinas will eventually argue, moreover, that careful study of this awareness of others will show that we perceive them as, in some sense, transcending our ability to comprehend them. Much more needs to be said, of course, about the details of his argument to make these claims compelling. The point I want to stress for now is simply that, if I am right in my interpretation of his murder argument, it stands to reason that his argument should be read as the centerpiece of *Totality and Infinity*.

Again, Levinas indicates in the opening section of *Totality and Infinity* that his aim is to prove the existence of something like mind-independent reality. To fulfill this goal, he needs to establish both (1) that we experience something in our lives such that it cannot be fully comprehended and (2) that the reality of this transcendent "something" is incontestable. There are several places in *Totality and Infinity* where he remarks on the first of these two points, but his murder argument is the only place where he argues both that acknowledging another person's humanity requires us to view her as transcendent *and* that we cannot help but acknowledge her in this way. For it is only with his murder argument that Levinas is concerned to establish that one cannot help but recognize other persons as possessing an inalienable dignity. There seem to be grounds, then, for attributing a special significance to his murder argument. It appears at key moments in what are arguably some of the most important texts of the most important period in Levinas's career. However, it also seems that his murder argument is presented in *Totality and Infinity* as the solution to the main problem of this work.

2. What's So Special about Murder?

Levinas argues that it is impossible to murder another person in "Is Ontology Fundamental?" and *Totality and Infinity*. However, a caveat is in order here. Levinas does not think it is impossible to *kill* another person. He grants that "the mere extermination of living beings" is possible.[8] So, he has something more elaborate in mind by murder than mere killing. What is so special about murder?

Murder is, first, a more deliberate act. I can accidentally kill someone, but I cannot accidentally murder someone. One needs to show that I killed my victim with malicious intent to establish that I am guilty of murder. Murder also differs from other acts of aggression in its exclusive focus on terminating the victim's life. Finally, murder seems to be directed at a specific class of entities. I can destroy my pencil by snapping it between my fingers, but I cannot murder it no matter how hard I try. This might seem trivially true: if murder aims at destroying *life*, then I can only murder *living things*. However, there is a deeper point to be noticed here. Suppose I ask the following: Can I murder animals? Some vegetarians argue that eating animals amounts to murder. However, their arguments usually involve attributing personlike qualities to animals; "meat is murder" because animals are sufficiently close to being personlike to make eating them as inexcusable as deliberately killing another person. So, it seems that we reserve the term *murder* for acts of aggression that aim at ending the existence of a being recognized to possess what I will call *humanity*—acts that deliberately aim at destroying a being recognized to possess whatever it is about us that makes us morally exceptional.

An act is murderous, then, if and only if the agent who performs it deliberately and exclusively intends by it to end the life of a being she recognizes to be a source of humanity. There is evidence that Levinas upholds this definition. He distinguishes murder from killing in several places. He also distinguishes it from other forms of aggression on the basis of its exclusive focus on "negation" rather than "domination."[9] Finally, he observes that murder is exclusively directed at other persons. He writes in "Is Ontology Fundamental?" and *Totality and Infinity* that "the human other (*autrui*) is the sole being I can wish to kill."[10] He also speaks of the other person's face as a "temptation" to murder.[11] His point in referring to the other's face as a temptation to murder is to stress that murder is a deliberate attempt to destroy not just any entity but one the murderer recognizes to embody moral value. The other's face tempts murder in the same way that sacredness tempts profanity. You cannot profane in a world devoid of the sacred. Similarly, there is a level of maliciousness that becomes live only when I recognize that I share the world with beings that are morally exceptional. I can be aggressive in a world of sticks and stones, but I cannot be perversely cruel without other people.

3. Murder's Impossibility

So, why does Levinas think that murder is impossible? I act murderously, again, if and only if I deliberately and exclusively try to kill a being I recognize to embody moral value. Levinas's claim is that recognizing someone *in this way* precludes deliberately harming her. If I self-consciously try to hurt someone, then I must be seeing her as something less precious, less awe inspiring than an embodiment of moral value.

Levinas's argument turns on an observation about the normativity of the ethical. It is widely accepted that ethical categories purport to be normative; to recognize an act *as ethical* means one recognizes it *as an act one should perform*. Suppose I am grappling with the question of the justness of going to war with another country; it may be possible to coerce me to *act* as if I held a certain view; I may feel so cowed by enthusiasm for the war that I hide my opposition. However, it is not clear that I can be coerced into *judging* it just. There is an important sense in which I need to reach this view myself. This is not to say it is wholly subjective—that any choice I make is right simply in virtue of my choosing it. The point is that seeing it *as right* involves recognizing it as exemplifying authority, as exemplifying what I *ought to do*, and I cannot be coerced into seeing it in this light. Instead, I need to explain my reasons for feeling reluctant about the war, submit them to your scrutiny, hear your reasons, and make a judgment about what ought to be done. Notice I have not said anything about the flavor of this normativity. I have not said if it pushes or pulls, if it is a carrot or a stick, a Kantian *you ought to do x* or a consequentialist *it would be better if you facilitated x*. All I have said is that ethical evaluations purport to be normative and thus to recognize an act as ethical means one recognizes it as normative. That is, to recognize an act as ethical means that one recognizes it as instantiating what one ought to do. Levinas thinks that the murderer's desire can be shown to be self-defeating if we grant this minimal assumption.

The murderer wants to see her victim, again, as embodying moral value, for otherwise she is not murdering but merely killing.[12] However, acknowledging her victim as embodying moral value requires her to see her as, to borrow Christine Korsgaard's phrase, a source of normativity. Levinas often characterizes this way of apprehending the human other in terms of inviolability. To apprehend the human other as a source of moral value is to apprehend her as a living embodiment of the biblical prescription against murder. "The face is what one cannot kill," he writes, "or at least it is that whose *meaning* consists in saying: *thou shalt not kill*."[13] Yet the murderer also wants to "negate" her victim; she wants to exert her power over her victim. However, this response would seem to be barred because she is supposed to view her victim as exemplifying authority. Thus Levinas notes that the "expression the face introduces into the world does not defy the feebleness of my powers but my very possibility for possibility" or "ability to be able" (*mon pouvoir de pouvoir*).[14] It is impossible to murder the other, that is, not because she is more powerful than I am, but because *if* I try to exert my power over her, then I *cannot* be viewing her as a human other, as a fount of normativity, in the first place.

Levinas's argument can be compared to one version of Anselm's ontological argument.[15] Anselm famously tries to show in his *Proslogium* that he can deduce God's existence from the atheist's denial of God. The atheist claims to deny the reality of a being *than-which-no-greater-can-be-conceived.* Anselm invites her to consider a being *than-which-no-greater-can-be-conceived* whose *nonexistence is inconceivable.* Surely, this second being is greater than the one she initially conceives when she denies God; a God whose nonexistence is *unfathomable* is more impressive than one who can be doubted. However, this means that the atheist denied the wrong thing. If she wants to deny *God,* she needs to deny the existence of a being whose nonexistence is inconceivable. This cannot be done: thus atheism is beset by an "irreconcilable contradiction."[16] Levinas gives the ethical equivalent of this argument. The murderer wants to negate not just any entity but one that embodies what I am calling "humanity." She wants to negate something she recognizes as embodying moral value. Recognizing her victim as a fount of such value requires her to see her as instantiating normativity. Yet if she perceives her victim as normative, she must be perceiving her as something she recognizes she *ought not* harm. Thus Levinas speaks of murder as an attempt "to exercise power over what escapes power."[17] The murderer, like the atheist, is beset by a contradiction. She wants, as it were, to profane something sacred even while she sees it *as sacred,* but she can profane the sacred only if she abandons this perspective, if she sees it as something less than inviolable, less than sacred.

I will consider objections to this argument in the next section. For now, let me just clarify its overall tone. The tension Levinas finds in murder resembles what Avishai Margalit calls "the paradox of humiliation."[18] Margalit argues that humiliating behavior is contradictory. Suppose I want to humiliate you: I want to force you to see yourself *as subhuman,* as something that does not merit my respect. However, I can achieve this end only if I effectively grant you your personhood. I can get you to *think* of yourself as subhuman only if I grant that you are a reflective being capable of moral evaluation. Frederick Douglass explores this tension with dazzling subtlety, I find, in his autobiography, *Narrative of the Life of Frederick Douglass, an American Slave*. Here is one example. Douglass recalls seeing slaves whipped as a child when the horses they used in fieldwork refused to plow. It puzzled him that his fellow slaves were singled out for punishment despite the fact that the same rhetoric was used to describe both slaves and horses. His master refused to whip his horses because he knew beating them would not make them work harder and might damage a valued commodity; the fact that he treated them well showed he saw them as mere commodities. Similarly, his exaggerated punishment of his slaves showed he could not have thought of them as subhuman. His abuse had to be hyperbolic to convince them, and perhaps himself, that they were subhuman, that they were mere objects rather than persons. However, the very theatricality of his punishment revealed that he could not have genuinely believed this was the case.[19]

Efforts to violate humanity have, then, a theatricality that reveals them to be contradictory. It is as if the cruelty we direct towards others needs to be hyperbolic precisely because it takes place against a background awareness of their humanity. Our cruelty needs to be exaggerated so that we can, as it were, assure ourselves that we are not moved by our victims' humanity.[20] However, the very fact that it is exaggerated betrays the extent to which their humanity moves us.

Levinas sees murder as exemplifying this conflicted drive to violate the human in its most unvarnished form. The murderer exemplifies the sheer urge to subject humanity to force with no other goal than crushing it. Levinas's claim is that this impulse is self-defeating. For such radical evil makes sense only against a background concept of the good as a source of value whose authority precludes its violation. The fantasy of murder reveals more, we might say, about our anxiety over the good than our longing for evil. It feigns the hope of our proving able to retain poise before a source of value whose nature consists in compelling us to recognize an authority in our lives greater than our own. In this way, murder might be said to represent a certain fantasy of isolation or insulation: the fantasy of being able to live untouched by others in a world where their suffering cannot help but rend our souls.[21]

4. Objections, Replies, and Indirect Angles: Or, Evil

I can envision two sets of objections to my reading of Levinas. One would come from continental philosophers who might feel I have misrepresented him. Another would come from analytic philosophers skeptical about the soundness of the argument I have attributed to him. Let me postpone defending the accuracy of my reading so I can focus on defending its soundness.

I can envision two objections to the argument I have attributed to Levinas. You might argue it is possible to fail to do things we know we should do and to do things we know we should not. I know I should pay my bills on time, but this does not stop me from sending them in long after they are due. So too, a murderer can recognize that she ought not harm her victim yet harm her anyway. You might also argue that Levinas's definition of murder is utopian. Murder might be impossible if murderers actually saw their victims as manifestations of moral value, but this is too good to be true. Do rapists really see their victims as persons rather than mere bodies or sources of pleasure? Did southern whites really see blacks as persons during slavery? Does not Raskolnikov, the archetypal murderer, say that his victim, the old pawnbroker, Alyona Ivanovna, is a mere "louse?"[22] A murderer might not deny, then, that her victim is human in the sense of being *Homo sapiens*. However, does she really see her victim as human in the sense of instantiating moral value?

Levinas would accept both criticisms. He would agree, first, that it is not *physically impossible* to act irrationally, but this observation is not profound and does not tell against his argument. I can recognize that you are making a sound disjunctive syllogism yet act contrary to your conclusion;[23] so too it is *physically possible* for murderers to kill their victims even if they recognize them to be morally valuable. Thus Levinas sometimes equates evil with Being or ontology.[24] His concept of ontology is, to be sure, a complex topic, part of his extensive critique of Heidegger's fundamental ontology. However, I would argue that his equating evil with ontology is intended, at a minimum, to stress that there is a gap between ethical and physical or metaphysical necessity that renders evil possible:

> Murder, it is true, is a banal fact: one can kill the Other; *the ethical exigency is not an ontological necessity*. The prohibition against killing does not render murder impossible, even if the authority of the prohibition is maintained in the bad conscience about the accomplished evil.[25]

Levinas would note, then, that his argument does not claim to show that murder is physically or ontologically or metaphysically impossible. It *does show* that murder is irrational or involves "bad conscience" insofar as the murderer purports to be able simultaneously to recognize her victim's humanity and violate her.

Levinas also would not be troubled by the observation that murderers typically view their victims as valueless or as subhuman or nonhuman. However, he would suggest that this insensitivity is telling: it shows that they cannot simultaneously recognize their victims' humanity and harm them. They *would* recognize a contradiction in their behavior if they were clearheaded; thus they need to resort to self-deception to blind themselves to their victims' humanity. Indeed, he would suggest that such self-deception is paradigmatic of evil. Levinas here resembles Plato when the latter claims in *Meno* that it is impossible to do evil knowingly and equates evil with ignorance. Levinas would be sympathetic to this claim, though he would not cast it in Plato's *eudaemonist* terms. Plato holds that evil results from our ignorance as to what acts conduce to living meaningful human lives. Levinas also thinks that we cannot knowingly do evil—that murder is impossible—and that evil must result from a form of ignorance. However, this ignorance is more a matter of my finding ways to blind myself to your preciousness, my feigned inattentiveness to what Levinas calls "the sacredness of the other's hunger."

Thus Levinas links violence's possibility, in one essay, with viewing the other's face from "an indirect angle." I cannot directly apprehend you as instantiating moral value, as embodying *Thou shall not kill*, and exert power over you, but there are ways to divert my attention. Levinas sees a special danger in the temptation to view others under the guise of abstractions or universals, token cases of, say, a biological species, sex, race, ethnicity, or member of an opposing political faction:

> [V]iolence consists in…ignoring the face of a being, avoiding the gaze, and catching sight of an angle whereby the *no* inscribed on a face by the very fact that it is a face becomes a hostile and submissive force.
>
> Violence is a way of acting on every thing…by approaching it from an indirect angle. Violence is a way of taking hold of a being by surprise, of taking hold of it in its absence, in what is not properly speaking it. The relationship with things, the domination of things, this way of being over them, consists in never approaching them in their individuality. The individuality of a thing…is in reality only accessible starting from generality, the universal… One grasps hold of a thing out of its concepts.[26]

Levinas's prose is a little opaque, but his basic aim is to contrast interpersonal relations with the way we relate to "things." He associates our "domination" of mere "things" with viewing them as instantiations of universals. I never truly relate to a thing in its individuality; I always perceive it as an instantiation of a universal, as an instance of a chair, a shape, a color, and so on. Violence is possible, Levinas suggests, because we can take this stance toward other persons. I can blur my sense of you as an immediate, utterly particular source of value by treating you as if you were a thing and fixating on properties you share with others. I can divert my recognition of you as an immediate source of value by viewing you as an instance of a more general type, just another *Homo sapiens*, Jew, Arab, black, woman, and so on.

Levinas would grant, then, that one can retrospectively recognize that one *virtually* did evil. Thus he speaks of murder as "the triumph of a power" exulting in "its defeat as power."[27] Strictly speaking, the murderer cannot perform the act she sets out to perform: she cannot deliberately lord her power over humanity. However, she can indulge in this illusion, indulge in the triumph of power unknowingly exulting in its defeat, by exploiting (1) the gap between ethical and ontological necessity, (2) viewing her victim as an abstraction, (3) harming her, and (4) recognizing in retrospect that she was able to destroy something human.

5. Realism with a (Real) Human Face: Ethical Metaphysics

What does Levinas establish through his murder argument? I suggested earlier that his murder argument plays a key role in *Totality and Infinity*. His goal in *Totality and Infinity*, as he explains in section 1, is to prove the existence of something like mind-independent reality. He wants to show that we experience things in our lives for which it is implausible to think we could have authored them ourselves using any philosophic system. His murder argument is intended to establish this point by proving that (1) we cannot help but recognize other persons as having a certain dignity yet (2) regarding them in this light requires us to view them as, in some sense, transcendent. I find it helpful in explaining how his murder argument entails these two points to invoke the language of moral realism and to suggest that Levinas can be read as using his argument to defend and identify the consequences of an extremely thoroughgoing moral realism.

Now, I want to be careful in making this suggestion. "Moral realism" is a loaded term; it has been linked in analytic moral philosophy with any number of different claims.[28] Levinas would accept some of the ideas associated with moral realism but would deny others. Nonetheless, I find it useful to use this vocabulary for two reasons. First, he seems to be interested in drawing realist conclusions based on his murder argument. His argument is intended to show that morality has a certain objectivity or authority. His murderer is someone who thinks she does not have to recognize morality's authority. Levinas is concerned to document an element of bad faith in this assertion; even the most flagrant violation of morality confirms its authority. However, it also seems to me, second, that he reaches this realist conclusion by, as it were, showing that a thoroughgoing moral realism is self-justifying. His murder argument has a certain hypothetical structure to it. He begins with the murderer's claim to be able to acknowledge another person as possessing moral value, and he invites us to think through what it would mean truly to perceive another person in this light. What would it mean truly to perceive another as a raw manifestation of "oughtness?" Levinas finds in reflecting on the challenges involved in acknowledging such an entity that its reality is peculiarly immune to doubt.

I find it helpful in explicating this line of reasoning to compare his argument with J. L. Mackie's influential essay, "The Subjectivity of Values."[29] Mackie argues in his essay that moral realism is untenable. The realist holds that moral predicates like *good* or *right* refer to real properties of things. Mackie insists that anyone holding this position posits implausible entities. The realist posits intrinsically normative moral facts, facts that radiate sheer *to-be-pursuedness* or *not-to-be-doneness*.[30] However, such facts would have to be radically unlike anything else in nature, anything science has revealed to us, and thus it seems as if we would need a special cognitive faculty, a kind of moral spider sense, to have knowledge of them. Moral realists posit entities so metaphysically and epistemologically "queer" that they cannot be taken seriously.[31]

Levinas runs this argument in the reverse direction. Mackie's presentation of moral realism is fair: realists *do* posit moral facts that radiate sheer *to-be-pursuedness* or *not-to-be-doneness*. Thus Levinas speaks of "the face of the other" as a concrete embodiment of the biblical command against murder: "The face is what one cannot kill, or at least it is that whose *meaning* consists in saying: *thou shalt not kill*."[32] Levinas also seems to accept Mackie's claim that such facts must be metaphysically and epistemologically queer. Thus he associates ethics with a plane of experience that transcends the category of being[33] and stresses that the other cannot be reduced to an object of comprehension. So, he seems to be sympathetic to Mackie's worry that realism commits us to supernaturalism and intuitionism.[34] However, he would also have us note that perhaps the "queerest" feature of moral facts is that they cannot be denied. Precisely because they radiate sheer *to-be-pursuedness* or *not-to-be-doneness* or *thou-shalt-not-killness*, they *morally prohibit us* from doubting their authority and hence are what we might call "performatively immune" to coherent questioning.[35]

The murderer's inability self-consciously to kill her victim illustrates this immunity. The murderer wants to acknowledge her victim as a locus of what Levinas calls "ethical resistance" yet violate her. However, she cannot in good faith do both at once. To borrow Mackie's terms, she cannot recognize her victim as a *moral fact* because recognizing her in this guise would impel her to see her as sheer *not-to-be-violated-ness* and hence as something she is prohibited from violating. I can perhaps put Levinas's point this way: I am being disingenuous if I ask a drowning man to prove I have a moral responsibility to help him. This request is insincere because I am asking him to show me, in effect, that I should not be asking this question. For if I recognized him as morally entitled to lay claim on me, I would recognize it as irresponsible to hesitate to help or question his claim. So, there is a sense in which ethics is self-justifying. I can entertain doubts about the reality of things like tables, chairs, my hands, or my own existence, but I cannot gamely doubt the other's hunger without morally incriminating myself.[36] Moral facts insist on being taken seriously: precisely because they are *moral* facts, they *morally prohibit* us from doubting them. However, it follows that the murderer's claim to recognize her victim's moral status must be insincere. For if she had acknowledged her status as a moral fact, she would have had to acknowledge it as impermissible to violate her victim.

We can read Levinas, then, as advancing an extremely thoroughgoing form of moral realism. He invites us to assume provisionally the existence of genuine moral facts, facts that blur the entire fact/value distinction. Suppose for the sake of argument that there exist entities such that the very act of acknowledging them makes practical demands on one's life—obligates one to, say, treat them with a certain respect. Levinas finds in reflecting on the challenges involved in acknowledging the presence of such entities that their reality is oddly immune to doubt. He is concerned to show, then, that a thoroughgoing moral realism is self-justifying. I find it difficult, though, to give a concise description of this realism. Again, the phrase "moral realism" is vague, and it has been used to advance a wide range of claims. However, Levinas's realism is especially hard to describe since he would repudiate many of the claims advanced by analytic moral realists precisely for not being *realist enough*.

Consider his antinaturalism. Most recent realists defend some form of naturalism.[37] The common wisdom seems to be that texts such as Mackie's "Subjectivity of Values," Gilbert Harman's *The Nature of Morality*, and Bernard Williams's *Ethics and the Limits of Philosophy*[38] have discredited realisms typical of rational intuitionism, realism to the effect that there exists "a realm of irreducible, observer-independent, and nonnatural ethical properties…that can be known through a purely intellectual perception."[39] The only respectable realism is one that can establish that moral facts are real because they are natural facts.

Levinas opposes this view. His thinking here runs entirely against the grain of recent analytic theorizing. The consensus view in analytic philosophy seems to be that moral realists have a harder time defending themselves than scientific realists; it is easier to be a skeptic about moral facts than about unobservable entities like gluons.[40] Levinas would dismiss this claim on the grounds that likening ethical certainty to scientific certainty *understates* its authority. One could imagine him arguing the following. Suppose science shows we are amoral, self-interested beasts. It is conceivable that you would persist in treating others as objects of unconditional moral respect because you would find it degrading to view them as mere beasts. You might find yourself, that is, *morally obligated* to dismiss science's findings; the normative content of the moral *ought* would be so final that it would trump what you are epistemologically justified in claiming to know to be the case.

This view may sound counterintuitive, but Levinas's perspective is faithful to the phenomenology of moral experience. Consider the resolve that defines us in the face of moral catastrophes. My life strikes me as small in the face of such catastrophes; I find myself moved to set aside my personal projects to attend to the greater good of others. I usually experience the world as a constellation of facts of varying degrees of recalcitrance and passivity, which I can mobilize in the service of my desires or life goals. However, this perspective shifts when I recognize myself as responsible for another: I feel as if *I am a passive fact* mobilized by an external authority. Moreover, this authority invests my life with certainty I cannot give it myself. Suppose I do not personally hold an opinion about affirmative action; I can see arguments for and against using race and sex as criteria in hiring procedures. My perspective changes in the face of the extreme poverty and unemployment in urban African American communities. Indecision is no longer licit. I recognize that something unconditionally wrong is happening, and *this realization* impels me to make judgments about what is just that I must uphold *as objective*. This is not to say it will be obvious what judgments I should make; the point is simply that I apprehend the abuse of others as an unconditional mandate to require certainty of myself in my convictions.

Levinas's realism is thus both familiar and heterodox. It is familiar in that his basic aim is to remain true to what is sometimes called the "finality of the moral ought."[41] He is struck by the urgency of ethical considerations, the way we experience them as overriding other claims on our attention, as unconditionally significant. His ambition is to develop a comprehensive philosophy grounded in this insight, what one commentator usefully calls an "ethical metaphysics."[42] His realism is heterodox, though, in that he is willing, in the name of this goal, to accept claims others would find implausible. Indeed, his differences from what typically gets associated with realism run even deeper than I have suggested. Realists usually posit moral facts to give a metaphysical justification for moral judgments. We make judgments that purport to be objective in our outrage at the world's injustices and then face the task of explaining how they can be objective. The realist's solution is to hold that they refer to real properties of things. However, this solution can seem *ad hoc*, as if the realist were inventing a metaphysic solely to excuse judgments she is duty bound to make.[43] Levinas invites us to dispense with this pretense. Ethical responsibility is exceptional precisely because it is unconditional, and because it is unconditional it directly imposes its authority onto us.

Levinas would have us abandon the idea, then, that ethics needs metaphysical foundations and instead to ascribe a foundational potential to it insofar as the normative content of the ethical relationship is so overriding that it directly imposes its reality onto us. He in a sense reverses the realist's strategy. The realist usually tries to show how moral judgments can fit within a respectable metaphysical picture of the world. By contrast, Levinas notes that ethical considerations do not stand in need of metaphysical corroboration because they are overriding, and, consequently, we should instead let them guide our metaphysical reflections. The task is not to get ethics to fit with metaphysics but to infer our metaphysics from ethics, to elaborate an "ethical metaphysics."

6. Why Is It Hard to Talk about Moral Facts?

Levinas's moral realism differs from the realisms typically offered in analytic philosophy in another key way. I noted earlier that Levinas wants to show in *Totality and Infinity* that other persons (*autrui*) are transcendent or absolutely other. He wants to show, that is, that the human other is both indisputably real yet not subject to full comprehension. I have so far concentrated on how his murder argument supports the former claim. Let me turn to the latter. The idea that the human other cannot be comprehended can sound as if it should contradict my reading of Levinas as a realist. We typically think in analytic philosophy that you should be able explain why you are justified in claiming to *know* that moral facts exist if you are going to call yourself a realist. Again, one of Mackie's complaints is that realists cannot answer this justificatory question since they posit "epistemologically queer" entities. Levinas's stance on this challenge can seem especially bad. He seems to be barred from formulating *any* response since he openly says that we cannot comprehend the human other. He seems to be claiming, in short, that moral facts are indubitable yet unknowable.

He insists on this point, though, not in spite of his realism *but because of it*. Levinas thinks we perceive a source of unconditional value in each person, but notice that apprehending others in this light will necessarily be a peculiar event. I can give a description of an everyday entity by listing off its properties, but faithfully reporting on my apprehension of something that, in its very appearance, sets practical constraints on me or commands responses from me will be different. My report will not be a *description* so much as a voicing of what *should be said.* A morbid analogy will shed light on this last claim. Levinas depicts moral deliberation as similar to giving a eulogy. A eulogy can sound like a description of a person's life, yet it is also an ethical gesture. Suppose I am asked to give a eulogy for a friend. My task is to find words that honor her. I will reflect on her life in preparing my speech. However, my aim is not simply to catalog things she accomplished; it is to discriminate moments in her life where I am awed by her. The words I offer are ones I find myself impelled to give in the face of her brilliance. They are not so much descriptions as my voicing what *ought* to be said.

Levinas takes moral deliberation to involve a similar process. I apprehend the vulnerability of others as an unconditional command to enact a fitting ethical response to it in my life. There is a danger, though, in depicting the choices I make by way of response as if they involved acts of comprehension. For this invites you to think of them as neutral reports on facts rather than ethical acts unto themselves.

Along these lines, Levinas suggests in several places that ethics and comprehension are not facts "on the same level." He was once asked after giving a paper if recognizing oneself as ethically responsible for another necessarily precludes comprehending this other person. Here is his reply: "Ethics and comprehension aren't on the same level. I substitute for comprehension not other relations, which would be *incomprehensions*, but rather that by which the comprehension of another alone begins to count for an I…"[44] Levinas's suggestion seems to be that our relationship to other persons is always, first and foremost, ethical in character. It is not that in others I discover things I cannot fathom, "incomprehensions." Rather, my responsibilities to them always supersede any other relations that hold between us, and it is potentially misleading, therefore, to characterize this ethical relation as if it were akin to other acts of cognition. Levinas makes a similar observation in *Totality and Infinity*: "The sense of our whole effort lies in affirming not that the Other forever escapes knowing, but that there is no meaning in speaking here of knowing or ignorance, for justice…is nowise, as one would like, a noesis correlative of a noema."[45] Levinas denies that in others we uncover something that "forever escapes knowing." His claim is rather that our relationship to others invariably has a practical dimension to it. The person whom I see begging for change on the street is not a neutral fact, something about which I may be in a state of ignorance, like, for example, my current ignorance as to the square root of 947. She is a being that, first and foremost, places practical demands on me, requires me to change my life. To return to the language of moral realism, a moral fact is not something one acknowledges through words but through deeds, and it is for this reason that Levinas uses the language of transcendence to discuss interpersonal relations.

7. Conclusion: Questioning the Ineffability Hypothesis

Let me conclude by returning to the question of whether my reading of Levinas is accurate. My analysis of Levinas's murder argument differs from others in the secondary literature. So, let me close by defending its accuracy and, more importantly, by saying something about the role I see it playing in my larger argument in this book.

Commentators sometimes explain Levinas's murder argument by appealing to his notion of "the absolutely other." Levinas proclaims early in *Totality and Infinity* that the human other (*autrui*) is absolutely other (*l'absolument autre*). Some interpret his murder argument as turning on this identification of *autre* with *autrui*. The murderer wants to violate the human other but can exert power only over what she can see, feel, comprehend, or act upon. However, if the human other is absolutely other, if she is transcendent, then she must be beyond the murderer's grasp. Thus Jill Robbins gives this exegesis of Levinas's murder argument, in *Altered Reading*:

> Murder wants to kill the other, who is (by definition) beyond the sensible. Yet in murdering the other, it arrives only at the sensible.

> In this way, murder always misses its mark. No doubt it effects an annihilation of the other in his being. But it thereby misses the genuine alterity of the other, namely, that which in him goes beyond the sensible.[46]

Robbins presents murder as impossible for roughly the same reasons why it is impossible to harass a Kantian thing-in-itself. You cannot harm a thing-in-itself, a thing as it stands apart from all possible experience, because it is, by definition, not something you can in any way contact. Similarly, Robbins reads Levinas as arguing that the human other cannot be murdered because it is beyond all direct experience.[47]

Let me confine myself to two criticisms. Notice that Robbins's analysis cannot explain why Levinas says that the other's face paralyzes the murderer's very "possibility for possibility" (*pouvoir de pouvoir*).[48] Her reading presents the murderer as unable to perform *a specific act*—unable directly to affect the other—yet Levinas presents the other person as undermining *efficaciousness as such*. Robbins cannot explain this detail. She presents the human other as akin to Kant's notion of the thing-in-itself, but there is nothing in the idea of a nonsensible entity that is incompatible with efficaciousness.[49] The fact that I cannot have any awareness of some object, such as my pencil, as it exists apart from all possible experiences of it, does not imply that I cannot perform any number of acts, such as getting up and walking around my chair twice before writing the next sentence.

My analysis of Levinas's argument makes better sense of his claim that the other's face undermines the murderer's very "possibility for possibility." The murderer claims to recognize her victim's moral status, but if this were the case, she would need to recognize in her victim a source of unconditional authority. So, the affront to the murderer goes deeper, on my reading. It is not simply that she discovers something she cannot destroy in the other; she is forced to acknowledge an authority she cannot help but recognize as imposing legitimate constraints on her will.

My larger concern, though, is that Robbins takes the claim that the human other is absolutely other to be uncontroversial. She presents Levinas's argument as relying on this observation: *the murderer cannot destroy her victim because the other is transcendent, hence beyond all possible interaction.* However, why should we accept the claim that the other is transcendent in the first place? My sense is that Robbins takes this claim to be established elsewhere in Levinas, or, what is more likely, finds it so central to his thinking that it can be assumed for the purposes of elucidating his argument. Yet a close reading of *Totality and Infinity* suggests that his murder argument is intended to prove this very point.[50] His argument is also apt to seem unconvincing in the absence of a defense of this claim. Murder might be impossible if there was an insensible, transcendent *something* at the core of each person, something on the order of a noumenal soul, but why should we think there is anything transcendent about others in the first place?[51] So, I differ in the following way. Robbins seems to think that Levinas's claim about the other's transcendence is uncontroversial. Alternatively, she may think that it follows from his critique of philosophic inquiry. I insist that his critique of philosophy cannot establish the claim that other persons are transcendent; it cannot because merely showing that the philosophers have found it difficult to conceptualize transcendence would not be adequate to show (1) that it exists or, more strongly, (2) that it is made manifest in compassion. Levinas needs a more focused argument to establish these points. His murder argument is this argument.

Robbins's error is symptomatic, however, of a broader problem in Levinas scholarship. There is a tendency to assume that Levinas's claims about the other's transcendence are preinterpretative or uncontroversial. His claim that each life is miraculous does not strike us as contentious; it rings true, perhaps because interpersonal relations are so rich, and what draws Levinas's readers to him, arguably, is the boldness with which he voices this conviction. The problem is that we overlook what he hopes to show us about *why* we find others miraculous, and the corresponding danger is that we inappropriately invoke this claim to do explanatory work elsewhere in his philosophy.

There is a special temptation, I find, to treat his claim about the other's ineffability as the foundational idea of his philosophy and to specify his ethics based on it.[52] Let me give an example. I quote from Axel Honneth's

"The Other of Justice": "In this situation of an unintended deprivation of liberty, there is what Levinas believes to be an inner-worldly experience of infinity: *My-vis-à-vis is a person who, in his or her unrepresentable individuality, is so incalculable that I am presented with the demand to render help infinitely.*"[53] Honneth presents Levinas as claiming that we have a duty to care for others because they are "incalculable." His *ethical* claim about infinite responsibility is presented as a consequence of his *epistemic* claims about the other's ineffability. Honneth's aim in his essay, moreover, is not to argue for this point, but to elucidate Levinas's challenge to Kantian ethical theories. It is revealing, however, that Honneth sees no need to explain or defend the argument that we have an "infinite obligation" to help the other because the other is "incalculable." That is, the assumption that Levinas makes this argument is so central to the secondary literature that Honneth sees no need to explain or to defend it. Instead, he feels comfortable simply presenting it as an innocent paraphrase of one of the main claims of Levinas's philosophy, one so central to Levinas's thinking that it can be assumed in the course of discussing Levinas's stance on Kantian ethical theories.

I want to question this innocence of this interpretative assumption. For this way of reading Levinas reverses the arc of his thinking. His claim *is not* that others have an infinite ethical claim on us because they are "incalculable." His epistemic claims about the incalculability of others are presented as consequences of the finality of ethical responsibility. Here I find it useful to recall his claim that responsibility and comprehension are not "on the same level." We saw earlier that he denies that there is something about other persons that "forever escapes knowing." His point is rather that our primary relationship to another person is not best described in terms of knowing or comprehension in the first place: it is first and foremost a relationship to something that makes practical demands on us, that requires change in our lives. So, I read Levinas's claims about the other's transcendence differently than do others. I do not deny that he uses the language of transcendence and infinitude to discuss face-to-face encounters—but he uses this language to stress that our relationship to others invariably has a practical aspect to it. To acknowledge humanity is to recognize something that requires a practical response on our part, and the true test of whether we recognize this fact is not borne out in claims to know it but in deeds performed in response to it.

Let me close by observing that we find grounds for rejecting the ineffability hypothesis when we read Levinas this way. The ineffability hypothesis holds that he cannot specify a normative ethics without violating his claims about the other's transcendence. His claim that the other is transcendent is not, however, the foundational idea of his philosophy. It is a consequence of his remarks on the finality of moral responsibility. It is wrong to conclude, therefore, that he thinks that the human other is incomprehensible, as it were, *simpliciter*. He has a story to tell about why acknowledging others as ethically precious requires us to view them as *in some sense* eluding comprehension.

It is not clear that the ineffability hypothesis survives the telling of this story. For Levinas's claim does not seem to be that we discover anything in others that is, strictly speaking, indescribable. His point seems to be that any description we offer by way of testament to our experience of them is subject to ethical considerations that do not arise in relation to neutral facts. I may be speaking falsely if I claim that Minneapolis is the capital of Minnesota, but I am not saying anything libelous. These kinds of considerations are unavoidable, though, in describing our awareness of others. So, one can imagine Levinas concluding, *pace* the ineffability hypothesis, that there is nothing wrong with feeling moved to develop a rich normative ethics as a result of one's face-to-face encounters with the other. He would not worry, that is, that we thereby misrepresent the other's ineffability. His concern would simply be that we retain a proper appreciation for the way in which even this act constitutes an ethical gesture on our part.

Endnotes

1. Levinas, "Is Ontology Fundamental?" in *Basic Philosophical Writings*, ed. Adriaan T. Peperzak, Simon Critchley, and Robert Bernasconi (Bloomington, IN: Indiana University Press, 1996), 6. See also Levinas, *Totality and Infinity*, 50.
2. See Levinas, "Transcendence and Height," 20.
3. Levinas, *Totality and Infinity*, 43.
4. See, for example, Levinas, "Transcendence and Height," 16; *Totality and Infinity*, 236.
5. See Levinas, *On Escape*, trans. Bettina Bergo (Stanford: Stanford University Press, 2003), 73.
6. Levinas, *Totality and Infinity*, 50.
7. Levinas, *Totality and Infinity*, 41.
8. See Levinas, *Totality and Infinity*, 199. See also "Is Ontology Fundamental?," 9: "I can, for sure, in killing attain a goal; I can kill as I hunt or slaughter or as I fell trees."
9. Levinas, *Totality and Infinity*, 198.
10. See ibid., 198; "Is Ontology Fundamental?," 9. Levinas's reference to killing in this quotation may seem out of place given his efforts to distinguish murder from mere killing. Levinas does distinguish these two acts, but he is not altogether rigorous in his terminology. He sometimes cites the biblical command against killing, for example, in discussing murder, suggesting that this command, *Thou shall not kill!*, is made manifest in the face of the other. Levinas here uses the term "kill," but it is fairly clear that he has in mind what he calls "murder" in his more guarded moments.
11. Levinas, *Totality and Infinity*, 194.
12. This claim may seem forced. It perhaps helps to suggest that Levinas has *Crime and Punishment* in the back of his mind. Levinas was an avid and subtle reader of Dostoevsky, and he frequently quotes from Dostoevsky's novels to illustrate his philosophy. He credits Dostoevsky as being "the first philosopher" he studied. See Levinas, *Ethics and Infinity*, trans. Richard Cohen (Pittsburgh: Duquesne University Press, 1985), 22. He was fond of quoting Father Zosima's dictum in the *Brothers Karamazov* to summarize his philosophy: *Everyone is responsible for everyone else, and I above all.* (See Levinas, *Ethics and Infinity*, 98, 101; *Otherwise than Being Or Beyond Essence*, 146; *Entre Nous*, trans. Michael B. Smith and Barbara

Harshav (New York: Columbia University Press, 1998), 105, 107, 168, for a few of Levinas's many references to the *Brothers Karamazov*.) Finally, he occasionally makes reference to *Crime and Punishment* in his work. See Levinas, "Meaning and Sense," in *Basic Philosophical Writings*, edited by Adriaan T. Peperzak, Simon Critchley, and Robert Bernasconi (Bloomington, IN: Indiana University Press, 1996), 52.

I find it unlikely that Levinas could have penned his murder argument without Raskolnikov in his thoughts. Raskolnikov exemplifies the qualities he identifies with murder. Raskolnikov wants to test his superiority by deliberately killing another person. His thought is that it will tell in favor of his superiority if he can perform an act that most people forbid themselves. Thus it is pivotal to his project that he recognize that the pawnbroker, Alyona Ivanovna, is morally unique, that she is not, as he sometimes says, a "louse." For if killing her is morally trivial, it will not tell in favor of his superiority. Raskolnikov exemplifies, then, the kind of act Levinas considers impossible. He wants knowingly to destroy something he recognizes as a legitimate object of moral respect. It will not be enough merely to *kill* Alyona. He needs to *murder* her; he needs to kill her in full knowledge of her status as a legitimate object of moral respect if he is to prove his superiority. Levinas thinks *this act* cannot be performed.

13. Levinas, *Ethics and Infinity*, 87. Again, Levinas uses the term "kill" in this passage, where he is quoting from the Bible, but it is clear from the context that he actually has murder in mind.
14. Levinas, *Totality and Infinity*, 198. I here translate Levinas's French differently from Alphonso Lingis. Lingis translates Levinas's *Mon pouvoir de pouvoir* as "ability for power." This translation is fine. However, I worry that the French word *pouvoir* has modal logic associations that are not obvious with the English word "power." *Pouvoir* can also be translated as "can" or "to be able," as in "I can swim" (*Je peux nager*). The English word "power" also conveys this idea of being able to do or accomplish something. However, it also has associations with might and force. I worry that this link between possibility and strength in the English word "power" may distort Levinas's point. That is, it makes it sound as though the face of the other paralyzes the murderer on a purely physical level—that it paralyzes the murder's strength or might, the murder's "power." Levinas seems, however, to be making a much stronger claim. The face of the other paralyzes the murderer's *efficaciousness as such*. That is, it exposes a limit on the murderer's very "possibility for possibility," "ability for ability," or "ability to be able." I will therefore translate Levinas's *pouvoir de pouvoir*

as "possibility for possibility," despite the awkwardness of this translation, so as to keep the modal logic associations of Levinas's argument clear for the reader.

15. There is some debate as to how many arguments Anselm gives in the *Proslogium*. I mean to compare Levinas with his argument in chapter III of the *Proslogium*—the argument in which he shows not simply that God exists but that God's nonexistence is inconceivable.
16. Anselm, "Proslogium," in *Basic Writings*, trans. S. N. Deane (La Salle, IL: Open Court, 1962), 8.
17. Levinas, *Totality and Infinity*, 198.
18. Margalit, *The Decent Society*, trans. Naomi Goldblum (Cambridge, MA: Harvard University Press, 1996), 108–112, 115.
19. See Douglass, *Narrative of the Life of Frederick Douglass* (New York: Penguin Books, 1986), 63.
20. There is a scene in the movie *The Thin Red Line* that illustrates this point. The soldiers in Charlie Company get caught in a rainstorm while they are recuperating on a beach. The camera lingers on Private Mazzi, who is crying in the rain. Prior to this moment he has been portrayed as the least sympathetic figure in his company. He is bloodthirsty: he tortures the Japanese prisoners before his company executes them and collects the teeth of his victims. He is shown crying over his bag of teeth in the beach scene. The scene is haunting because we realize his cruelty is a façade he has cultivated to convince himself he is not traumatized by the killings he has to commit, a strategy he uses to inoculate himself against his victims' and his own humanity. It is easier for him to think of himself as a bloodthirsty monster than to acknowledge that it pains him to have to kill.
21. Evidence for this last claim can be found in Levinas's remarks on the myth of Gyges. Plato famously uses the myth of Gyges and his invisible ring in the *Republic* to illustrate moral skepticism. Levinas also cites the myth of Gyges as illustrating the nature of immorality. See for example Levinas, "Truth of Disclosure and Truth of Testimony," in *Basic Philosophical Writings*, ed. Adriaan T. Peperzak, Simon Critchley, and Robert Bernasconi (Bloomington, IN: Indiana University Press, 1996), 104–105. Unlike Plato, though, he is interested in the way the myth links immorality with voyeurism. Plato uses the myth to show how someone might commit crimes without any fear of being punished. However, Levinas is more interested in the myth's literary details, the way Gyges's ring lets him enter into a one-sided relationship with other persons, lets him look on others without being judged or looked upon in return. There is the suggestion in Levinas,

then, that evil tempts us because it embodies a certain fantasy of being able to hide ourselves from others, to insulate or inoculate ourselves to the deep responsibility we feel for them.

22. Dostoevsky, *Crime and Punishment*, trans. Constance Garnett (New York: Random House, 1950), 374.
23. Suppose we go to see a movie. We have trouble finding a parking space and park illegally rather than be late for the film. Afterwards, I urge you to run back to the car because I am worried that we may have gotten a ticket. You note that we have either already gotten a ticket, in which case running will not help, or we have not, in which case we will be back in time to move the car. Your argument may be sound. However, its soundness does not *physically prevent* me from running back anyway. You would be entitled to accuse me of being irrational, though, if I accepted the soundness of your argument and ran back to the car in a fit of nervousness. I take it Levinas wants to identify a similar type of bad conscience at the root of evil.
24. See Levinas, "Philosophy, Justice, and Love," 114: "Evil is the order of being."
25. Levinas, *Ethics and Infinity*, 87, emphasis added.
26. Levinas, "Freedom and Command," in *Collected Philosophical Papers*, trans. Alphonso Lingis (Pittsburgh: Duquesne University Press, 1987), 19–20.
27. Levinas, "Is Ontology Fundamental?," 9.
28. Peter Railton identifies twelve distinct claims that are associated with moral realism in "Moral Realism," *The Philosophical Review* 95:2 (April 1986): 163–207. See also R. Jay Wallace's helpful discussion of moral realism in "Normativity and the Will," in *Normativity and the Will* (Cambridge: Cambridge University Press, 2006), 71–81. Michael Morgan considers whether Levinas should be associated with moral realism in *Discovering Levinas*, 228–230.
29. Mackie, "The Subjectivity of Values," in *Ethical Theory*, ed. James Rachels (Oxford: Oxford University Press, 1998), 76–102.
30. Mackie, "Subjectivity of Values," 95.
31. See ibid., 94–97, for his arguments against moral realism from metaphysical and epistemological queerness.
32. Levinas, *Ethics and Infinity*, 87.
33. Hence the title of Levinas's second major work: *Otherwise than Being; Or, Beyond Essence.*
34. Hilary Putnam notes Levinas's similarities with rational intuitionists in "Levinas and Judaism," in *Cambridge Companion to Levinas*, ed. Simon Critchley and Robert Bernasconi (Cambridge: Cambridge University Press,

2002), 33–62. *See esp.* p. 54. He credits Harry Frankfurt for first suggesting this similarity.

35. You might argue that this only bolsters Mackie's point. We *should not* be able to doubt moral facts if they exist. However, we *can*. So they cannot exist. However, notice how we are flirting with Raskolnikov's thoughts. Raskolnikov wants to show that good and evil are illusory by performing an act purported to be evil. He wants to *prove empirically* that "all is permitted," that morality is a sham. Yet what does killing Alyona show? Levinas admits it is *physically possible* to kill others, and Raskolnikov presumably knows this too. He does not kill Alyona to settle the eternal question: does hitting a *Homo sapiens* in the head with an axe damage its central nervous system? However, if Raskolnikov is not concerned to make *this point*, then what evidentiary work is Alyona's death supposed to do?

    It seems as if Raskolnikov wants to prove that he enjoys a certain freedom from the ethical—that he can stare down a moral fact. However, Levinas's point is that it is impossible, by hypothesis, to perform *this gesture*. I find it helpful to liken him to Anselm again. Anselm's atheist cannot doubt God because anything whose nonexistence he conceives will be too meager to count *as God*. The atheist, as it were, thinks about God, doubts God, and later realizes he could not have been thinking about God since God is supposed to be *absolutely great* and the thing he conceived would have been *greater* if he could not doubt it. So he starts over: thinks of God, doubts, realizes he doubts the wrong thing. And so on. The murderer is in a similar boat. Raskolnikov can show he is *physically able* to kill an old woman. However, *that* is not revealing. To prove more, though, he would need to show that he saw her *qua* moral fact even as he killed her. However, then he will always have to recognize in retrospect, like Anselm's atheist, that he could not have actually flaunted a moral fact, for he should not have been able to flaunt her authority had she been a moral fact, hence that the murder did not do its evidentiary work. The point, then, is that we cannot, by hypothesis, gainsay genuine moral facts, and so no gesture we could perform could be adequate to show that they do not exist.

36. It is worth noting that several psychological studies support Levinas on this point. Moll, de Oliveira-Souza, and Eslinger, for example, completed studies in which they measured the brain activity of subjects while evaluating moral sentences. They found that areas of the brain associated with emotional response were active when subjects were presented with moral sentences such as "You should break the law when necessary" but not with factual sentences such as "Stones are made of water." See

Moll, de Oliveira-Souza, and Eslinger, "Morals and the human brain: A working model," *Neuroreport* 14 (2003), 299–305. Similar results were found in Sanfey, Rilling, Aronson, Nystrom, and Cohen, "The neural basis of economic decision making in the ultimatum game," *Science* 300 (2003), 1755–1758; Berholz, Armony, Blair, and Dolan, "An fMRI study of intentional and unintentional (embarrassing) violations of social norms," *Brain* 125 (2002), 1696–1708; and Green, Sommerville, Nystrom, Darley, and Cohen, "An fMRI investigation of emotional engagement in moral judgment," *Science* 293 (2001), 2105–2108. There is, in short, an impressive body of literature in psychology which shows that we find it emotionally taxing to even contemplate violating moral norms—that we accord them a unique authority in our lives, one that distinguishes our contemplation of them from our contemplation of nonmoral propositions. See Prinz's "The Emotional Basis of Moral Judgments," *Philosophical Explorations* 9:1 (March 2006), 29–43, for an overview of this literature.

37. See Railton's "Moral Realism," and Sturgeon, "Moral Explanations," in *Morality, Reason, and Truth* (Totowa, NJ: Rowman & Littlefield, 1985), 49–78, reprinted in *Ethical Theory*, 198–227, for examples.
38. See chapter 1 of Gilbert Harman's *The Nature of Morality: An Introduction to Ethics* (Oxford University Press, 1977). See also Bernard Williams's *Ethics and the Limits of Philosophy* (Cambridge, MA: Harvard University Press, 1985), especially chapters 8 and 9.
39. I borrow this description of rational intuitionism from Darwall, *Philosophical Ethics* (Boulder, CO: Westview Press, 1998), 55, see also 49–54.
40. Both Harman and Williams argue against traditional moral theories by contrasting ethics with science. Mackie does not contrast ethical certainty with scientific certainty. However, he does dismiss the idea that there are objective values because he takes this view to be incompatible with the understanding of physical reality found in the sciences.
41. See for example Becker "The Finality of Moral Judgments," *Philosophical Review* 82 (1973): 364–371.
42. Levinas scholarship is indebted to Edith Wyschogrod for coining the phrase to describe Levinas's project. See Wyschogrod, *Emmanuel Levinas: the Problem of Ethical Metaphysics* (The Hague: Martinus Nijhoff, 1974).
43. Christine Korsgaard makes this criticism of realism in *The Sources of Normativity* (Cambridge: Cambridge University Press, 1996), 34–37.
44. Levinas, "Questions and Answers," in *Of God Who Comes to Mind*, trans. Bettina Bergo (Stanford: Stanford University Press, 1998), 99.

45. Levinas, *Totality and Infinity*, 89–90.
46. Robbins, *Altered Reading*, 92.
47. Bernhard Waldenfels develops a similar interpretation of the murder argument in "Levinas and the Face of the Other" *Cambridge Companion to Levinas*, ed. Simon Critchley and Robert Bernasconi (Cambridge: Cambridge University Press, 2002), 63–81. *See esp.* p. 70: "The other resists violence not as somebody belonging to the totality of beings, but as in infinite which is beyond all we can do to the other."
48. Levinas, *Totality and Infinity*, 198.
49. I would also level this criticism against Waldenfels's analysis of the murder argument in "Levinas and the Face of the Other." Waldenfels appeals to the idea that the other is "beyond all we can do to the other" in order to claim that the "otherness of the other manifests the impossibility of our own possibilities." However, again, there is nothing in the idea of a "thing in itself," a thing beyond all possible contact, that contradicts the notion of efficaciousness.
50. I gave evidence for this reading at the end of §1 of this chapter. To repeat: Levinas introduces the claim that the absolutely other (*autre*) is the human other (*autrui*) in section 1 of *Totality and Infinity*, where he abstractly reviews the philosophic problem he hopes to solve in *Totality and Infinity*, the problem of the absolutely other, and provisionally sketches his proposed solution. He proposes to show that the transcendent or "the idea of the infinite" is "concretized" in "the face of the other." He does not return to this claim until section 3. Levinas issues a promissory note, then, in section 1 eventually to explain how the other's face concretizes infinity. He opens section 3 by reminding us of this claim, with a chapter titled "Infinity and the Face." He here introduces the idea of ethical resistance to explain why the other cannot be comprehended. He gives his murder argument to illustrate ethical resistance in the following chapter, "Ethics and the Face." His murder argument is flagged, then, as his justification for the claim that the absolutely other is the human other.
51. I would add as a final point that her analysis can be applied equally well to anyone who accepts the problem of other minds. The existence of other minds is contentious insofar as mental properties are taken to be nonsensible. So, anyone who accepts the problem of other minds will find murder to be impossible for the reasons Robbins adduces. Her analysis, then, cannot distinguish Levinas's analysis of the problem of the other from more familiar versions. It purges his argument of what is most distinctive about it—its appeal to ethical properties as grounds for showing that the other is indisputably real.

52. Simon Critchley writes in his introduction to the *Cambridge Companion to Levinas*, for example, that Levinas's "big idea" is that there is a "relation with the other person that cannot be reduced to comprehension," 7–8.
53. Honneth, "The Other of Justice," in *Cambridge Companion to Habermas*, ed. Stephen K. White (Cambridge: Cambridge University Press, 1995), 289–324. *See esp.* p. 312, emphasis added.

Chapter 3

Making Sense of Levinas's Silence

But one must not be silent. We are not before an ineffable mystery.
—Levinas, "Questions and Answers"

1. Introduction

I made a pair of promises in chapter 1. I promised to show that Levinas *can* advance a normative ethics, and I promised to show that he *does* advance one, or, more guardedly, that hints of one can be found in his writings. Chapter 2 sought to fulfill the first promise by clarifying Levinas's murder argument in *Totality and Infinity*. His claim, we saw, is not that the human other is wholly ineffable, but that one's relationship to the other is primarily defined by moral responsibilities, and thus it differs in significant ways from the neutral, disengaged perspective that defines our relationship to objects of comprehension. The other is not something I *know* but something I *serve*. My further suggestion was that nothing in this view makes it impossible for Levinas to give a positive account of justice. I now want to return to my second promise. Levinas's critics have been so smitten by the idea that he *cannot* offer advice on how to be ethical that, I hope to show, they have ignored moments when he *does* offer advice.

My argument for this claim will proceed in two steps. For the first half of this chapter, I argue for it by citing passages where Levinas indicates that he is sympathetic to broadly Kantian, liberal ideals. In particular, I review passages where he (1) faults antihumanist philosophies for privileging abstract social processes over the rights of individuals as well as (2) passages where he expressly defends a Kantian "humanism of the other." In the second half, I take up this question: Why does Levinas give so little indication as to what ethics he favors if, as I claim, nothing prevents him from giving a positive account of justice? Does the ineffability hypothesis not supply us with a better explanation for this silence? Levinas has several pragmatic reasons, I show, for not elaborating a full, robust account of justice. Theories of justice are ultimately tools we use to assist ourselves in aiding others. Levinas is reluctant to elaborate a detailed theory of justice, or normative ethics, not because he finds it impossible to discuss whether one theory is superior to another, but because he worries that we tend to misuse our theories, to lose sight of the fact that they are tools designed to help us *act* for the sake of the other and, instead, to delude ourselves into thinking that we have fulfilled our responsibilities so long as we adhere to an accurate theory of the good.

These two steps may seem contradictory. On the one hand, I am saying that Levinas is a Kantian liberal. On the other hand, I am saying he is a pragmatist. Strictly speaking, these observations are not contradictory; one can uphold a pragmatic view in maintaining that moral theories are tools we use to guide interpersonal action, yet contend, on the basis of pragmatic considerations, that the norms found in Kantianism and liberalism represent our best set of tools. Indeed, Richard Rorty defends this position in several works.[1] This chapter's argument therefore is not contradictory. Yet it is somewhat fragmentary. My intent for now is not to elaborate a comprehensive explanation of how Levinas can give a positive account of justice, nor to assess in detail his compatibility with Kantianism, but simply to begin to motivate some skepticism in the reader as to whether his remarks on justice are as impoverished as his critics allege. My immediate goal is fairly modest: I am troubled by the allegation that Levinas does not say *anything* useful about ethics, and I want to show the he gives *some indication* as to what ethical perspective he favors.

I need to ask, then, for the reader's indulgence. I close this chapter by making an initial effort to unify my observations into a more general interpretation of Levinas's ethics. However, these remarks will need to remain somewhat provisional. I leave it to the next two chapters to fill in the details—to clarify, respectively, the extent to which Levinas's ethics is compatible with Kantian liberalism as well as how it is possible, on a pragmatic reading of him, to develop a positive account of justice.

Let me address one additional worry. Someone could object that it is a mistake to look to Levinas's *political* writings for insight into his *ethics*. Even if I am correct that he is sympathetic to liberal ideals, this tells us nothing about his ethics. After all, both utilitarians (Mill) and Kantians (Kant) have argued for liberalism, despite their adherence to rival normative ethical theories. So, too, establishing that Levinas is receptive to liberalism does not necessarily shed light on his ethics. One could argue that I am confusing ethics with politics. There is a substantial secondary literature on Levinas's concept of the political,[2] but this literature does not necessarily help us determine what ethics flows out of his remarks on the face-to-face encounter with the other, what approach to personal conduct is suggested by them.

There are a two points I would make to alleviate this worry. The first is that Levinas himself blurs the lines between ethics and politics. He associates "ethics" with the relation to the other, whereas he uses "justice" (*le droit*), particularly in later writings, to refer to the gamut of norms, policies, procedures, laws, and institutions we create to help us balance our responsibilities to multiple human others.[3] Justice refers, then, not only to normative ethical theories, but also to social and political philosophies.

There is not a sharp distinction between the two, ethics and politics, in Levinas.[4] Questions of justice can occur within the orbit of my personal life when, say, I am faced with the dilemma of whether to allocate greater time and resources to help those others who are closest to me, such as my spouse, my children, or my students, or whether, instead, I should devote them to strangers who are suffering under more extreme forms of poverty. Yet such questions of justice can also occur on a more sweeping, communal level. Should I vote in favor of politicians who sponsor protectionist economic policies that benefit workers in my home country? Or should I vote for politicians who maintain that we have a duty to bolster the economies of undeveloped countries? Levinas views our answers to such questions, all our answers, as enactments of justice, regardless of whether they involve issues we typically associate with ethics or issues we associate with politics.

The second point I would make is that it may not be inappropriate to look to Levinas's politics for clarification of his ethics given my somewhat limited aims. My concern is with critics who charge that Levinas's philosophy is flawed because he cannot offer *any* guidance on matters of practical significance. The charge against Levinas is fairly strong: his analysis of the relation to the other prohibits him from offering *any* advice on how respect for the other can be recorded in *any* set of policies, either in a personal code of ethics, a system of moral norms, or in more general social, political, and legal institutions. My intent is to dispute this point by showing that *some* discussion of precisely this issue can be found in his writings insofar as he expresses support for Kantian liberalism. Moreover, we will see that that his arguments for this political outlook tell us something of his moral sympathies. For example, he faults antihumanist political philosophies because they do not appreciate the extent to which they rely on a deeply humanist concern for the well-being of individuals. He also praises Kant's ethical tests because they model the calculations we are required to make in balancing responsibilities to multiple "human others." His arguments, then, provide clues as to his ethical perspective, such as that he seems to favor a broadly Kantian approach, a Kantianism that is not premised, however, on substantive claims about autonomy but on opposition to cruelty and a sense of the importance of fairness.

2. Humanism of the Other

Let me review some of my main claims so far. My aim has been to show that nothing in Levinas's thinking prevents him from developing a normative ethics. Many philosophers think that it would be contrary to the spirit of his philosophy for him to specify any ethical norms. His big idea is that the human other is absolutely transcendent, and he thinks that the great flaw of philosophers is that they have tended to disregard this ineffability. Yet Levinas would need to disregard it himself, critics charge, to develop a normative ethics. My aim has been to disabuse us of this worry by clarifying Levinas's argument for why the human other resists conceptualization.

The conviction that Levinas cannot advance a normative ethics is, however, all but unquestioned in the secondary literature, and a consequence of its ubiquity is that readers disregard moments when he indicates how his remarks on ethics might be put into practice. They are so smitten by the idea that he *cannot* offer advice that they overlook the advice he *does* give in his work. In particular, they overlook both moments where he *negatively* indicates his ethical stance by criticizing certain approaches in ethics and moments where he offers *positive* recommendations as to what sort of normative stance he favors.

Levinas frequently criticizes, for example, antihumanist perspectives. He faults ethical and political theories that focus on entities such as states, economic classes, historic processes, nations, or cultures as the primary datum of ethics and social and political philosophy. For example, there are passages in *Totality and Infinity* where he criticizes philosophies that, in the tradition of Hegel, treat persons as manifestations of broader historical processes. "When man truly approaches the Other," he notes, "he is uprooted from history."[5] Elsewhere in *Totality and Infinity* he associates one's encounter with the other with an "eschatological" break in history that requires us to "submit history as a whole to judgment."[6] Along these lines, Levinas sometimes expressed reservations about efforts to assimilate his ideas to Marxism. For example, he was once asked in an interview to comment on the reception of his work among South American liberation theologians, who had tried to wed his discussion of ethical responsibility to their brand of Marxist political critique. He explained that he admired the focus in liberation theology, and in Western or humanist Marxism, on issues of alienation and economic justice, but noted that he rejected the emphasis in Eastern or orthodox Marxism on "conquest."[7] He expressed reservations, that is, about the tendency among orthodox Marxists to view history as comprising the evolution of economic forces and to maintain that our actions are good, no matter how ruthless they might seem, if they facilitate the emergence of a classless society.

Levinas also wrote numerous essays in the sixties and early seventies in which he criticizes antihumanist philosophies.[8] "Ideology and Idealism" (1973) is a good example. In it, he questions the shift in French thought over the sixties toward avowedly antihumanist social and political philosophies. He alludes to various thinkers associated with structuralism and poststructuralism, such as Levi-Strauss, Althusser, Deleuze,[9] as well as Heidegger's 1947 "Letter on Humanism." Levinas focuses in his essay on the reception of Marxist ideology critique among his peers. Proponents of ideology critique accept Marx's idea that the economic structure of a society, its economic "base," conditions its culture and the conscious thoughts of its citizens, its ideological "superstructure." That is, what we consciously think about the world and ourselves is determined by the economic and material forces that ultimately define our lives. They argue, in turn, for a type of skepticism. We should doubt what is expressly asserted in any given philosophic theory since what occurs at the level of ideology inevitably camouflages its underlying socioeconomic processes. Levinas is intrigued by the acceptance of this critical strategy among his antihumanist peers, many of whom accept something like Marx's superstructure/base paradigm in claiming that persons are nothing more than nexuses of broader social, historical, or economic forces. Many also accept a similar skepticism, claiming that we should second-guess philosophic arguments since they appeal to our mistaken conception of ourselves as, first and foremost, rational agents rather than concatenations of biological drives and/or social processes.

For his part, Levinas tries to show that the ethical impetus behind ideology critique is at odds with his contemporaries' antihumanism. For all their talk about "the death of man" or "the death of the subject," French antihumanists accept a critical strategy that presupposes a humanistic respect for the dignity of human life.

Let me focus on Foucault in explaining Levinas's argument. Levinas does not expressly refer to Foucault in "Ideology and Idealism," but Foucault's *The Order of Things* (1966) was popular when he wrote this essay, and a key reason for its popularity was that it was taken to exemplify the antihumanist zeitgeist in French thinking at the time.[10] Levinas would have been familiar with its main theses and would have seen it as epitomizing the viewpoint he questions in his essay.

Foucault's ostensible goal in *The Order of Things* is to give a history of the disciplines of economics, zoology, and linguistics. His history is intended, though, to dramatize a philosophic point. One of his underlying ideas is that every historic period has a specific "epistemic space" that defines "what ideas can appear, what sciences can be constituted, what experiences can be reflected in philosophies."[11] This epistemic space does not necessarily correspond to any deep facts about the world but simply reflects what the current social, political, and scientific institutions define as reasonable. Foucault has, in turn, a therapeutic goal in his histories. In any historic period, certain ways of thinking will be stigmatized as irrational not because they are, in fact, faulty, but because the institutions that define that period stigmatize them as taboo. Foucault's aim is to demythologize some of the ideas that constrain our current thinking by alerting us to their contingency. Thus he sets out to show in *The History of Sexuality* that the very idea of sexuality is a recent historic invention, corresponding to no deep fact about our nature, and thus there is no reason why we should let our lives be ruled by this idea.

Foucault's target of criticism in *The Order of Thing* is the very idea of "humanity." We often think of ourselves as possessing some unique property that distinguishes us from other organisms. Foucault tries to show, however, that this sense of ourselves is a relatively recent invention, a fiction that has been foisted on us as a result of the rise of disciplines like economics, zoology, and linguistics. His aim is to expose the contingency of the very idea of human nature, and his hope is that doing so will enable us to realize that we are nothing more than concatenations of drives and social processes, that, as he famously puts it, "man is but a fold in language." Foucault's argument for this claim is less significant for my purposes than what Levinas does with it, or what he does with similar claims by other antihumanists. Levinas is interested in the tension between Foucault's therapeutic goal and his more hyperbolic antihumanist claims. On the one hand, Foucault denounces the very idea of humanity as a myth. On the other hand, he is keen to denounce this myth because he sees it as pernicious, as constraining the life possibilities available to us. For all his purported antihumanism, then, he seems to be a kind of eudaimonist, someone concerned to ask what it means to flourish as a human and to denounce institutions that inhibit us from flourishing. Thus Levinas observes in "Ideology and Idealism" that the "revolt against injustice" in antihumanist political critiques is ultimately grounded in a deep respect for what he calls "the right of the other person."[12]

Levinas also makes this point in *Otherwise than Being*. "Modern antihumanism," he notes, "is true over and beyond the reasons it gives itself. Its great intuition is to have abandoned the idea of the person as an end in itself. The Other is the end, and me, I am the hostage."[13] Levinas has some respect for antihumanist critiques; they identify an important flaw in traditional defenses of liberalism. We often defend liberal ideals by claiming that humans possess some special property, such as autonomy, that merits respect, and we often justify this claim by appealing to an immediate awareness of our own nature. I have an immediate sense of myself as, say, endowed with free will, and this leads me to conclude that I should respect others insofar as they are akin to me. Levinas finds this defense troubling because it is too easily misused; we too easily rationalize indifference toward others by convincing ourselves that they lack the properties that distinguish humanity. By contrast, Levinas argues for a "humanism of the other." He accepts the claim that each person has an inalienable dignity, but he thinks this claim needs to be grounded in the conviction that *other persons* are unconditionally valuable. I should accept liberal ideals, that is, not because I intuit in my own life some property that I then find reflected in others. Rather, the sheer preciousness of another's life, ultimately each and every other life, should awe me, and it is *their dignity* that should then lead me to accept liberal ideals.

"Modern antihumanism" is valuable, then, in that it curbs our temptation to seek out some general "human nature" to underwrite our respect for one another. Nonetheless, modern antihumanism is true "beyond the reasons it gives itself" in that its critiques are ultimately premised on a deep respect for humanity as it is disclosed to us in the respect we feel for others. As we saw earlier, antihumanists invoke a form of ideology critique because they are keen to denounce institutions that inhibit men and women from flourishing. Antihumanism is true, then, "beyond the reasons it gives itself" insofar as it rightly reflects the idea that humanity merits respect and, more specifically, that we reach this realization not by intuiting some general property disclosed to us in our own lives but rather through the respect others instill in us. Thus Levinas observes in the above quotation that antihumanism is compelling because it reflects our sense that one discovers something like an "end in itself" in "the Other."

Levinas gives some indication, then, as to what ethics he favors in the criticisms he levels against antihumanists. He is critical of broadly communitarian theories,[14] orthodox Marxism, and social and political philosophies founded on structuralist ideas. What he finds objectionable in these views is the way they privilege abstract historical or social processes over the welfare of individuals. So, as a first stab at describing his ethics, we might say that he accepts some of the basic ideals of liberalism. Each person has an irreducible, inalienable dignity that confers on her certain rights, and the main aim of our political and social institutions should be to protect these individual rights and secure the conditions needed for personal freedom.[15]

Levinas would perhaps also attach a special priority to institutions and policies that secure the conditions needed for basic physical well-being. I draw this conclusion for two reasons. First, his standard examples of ethical acts involve acts such as preventing cruelty, alleviating poverty, and feeding the hungry, "taking the bread from one's own mouth to nourish the hunger of another with one's own fasting."[16] We also saw, second, that he respects antihumanists for faulting arguments on behalf of liberalism that rely on substantive claims about human nature rather than the claim that in others we discern a source of unconditional value. So, while he accepts a liberal outlook as his starting point, he would perhaps attach greater significance to social welfare programs and issues involving economic justice. For example, John Stuart Mill also argues that our political institutions should aim to secure the conditions needed for personal freedom. He defends this claim, though, by claming that self-cultivation and individuality are supremely valuable. By contrast, Levinas seems to value liberal ideals because they reflect our sense that each life is inestimably valuable. So, whereas a liberal like Mill might attach a special significance to, say, our right to freedom of expression, Levinas would perhaps value more highly things like a robust national health care program, more compassionate welfare programs generally, and mandates that raise the national minimum wage. What matters, first and foremost, he might argue, is not that one has ample opportunities to cultivate oneself but that the basic needs of others are met.

3. Charity and Justice

One can also find evidence for this reading of Levinas in the *positive* claims he makes on behalf of certain ethical perspectives. I noted earlier that Levinas distinguishes between "ethics" and "justice." He reserves the term "ethics" for one's awareness of others as unconditionally valuable in the encounter with the other, and he uses "justice" to refer to systems of norms, principles, and policies we develop based on this awareness of others. He also associates justice with what he calls "the third party" or "the third" (*le tiers*):

> But I don't live in a world in which there is but one single "first comer," there is always a third party in the world: he or she is also my other…Hence, it is important to me to know which of the two

> takes precedence. Is the one the persecutor of the other? Must not human beings, who are incomparable, be compared? Thus justice, here, takes precedence over the taking upon oneself of the fate of the other.
>
> I must now judge, where before I was to assume responsibilities. Here is the birth of the theoretical; here the concern for justice is born.[17]

Levinas thinks that there are moments in our lives, face-to-face encounters, when we become keenly aware of another person as unconditionally valuable. Yet this awareness never occurs in a vacuum; it is always qualified by a corresponding awareness that we have an equally "infinite" responsibility to all other individuals as well. This perpetual presence of "the third party" introduces a need for considerations involving fairness and for general ethical principles. If the world contained only one other person and me, her claim on me would be absolute. However, I inevitably find myself bound up in a network of responsibilities and required to make difficult choices about how to apportion my time, attention, and resources. Thus I find myself faced with an obligation to develop fair and balanced ethical rules.

Levinas seems to be sympathetic, then, to the kind of model of ethical reflection found in Kant. Kant argues in his *Groundwork of the Metaphysic of Morals* that ethical rules are universal. When I claim that, say, sexual harassment is immoral, I am making a claim not only about certain forms of employment (it is immoral for businesspersons to harass secretaries) or certain sexes (it is immoral for men to harass women); I allege that harassment is wrong *whenever* and *wherever* it occurs. Moral judgments seem, that is, to identify universal laws, and Kant offers his various versions of the categorical imperative as tests we can use to assess whether our actions may express such laws. In particular, he often recommends that we use considerations involving publicity as a test for determining if we may be acting on the basis of a universal law. Can I envision myself standing before others and sharing the reasons for my actions with them? Can I reasonably expect them to accept them? If not, then I have reason to doubt that my act does, in fact, express a universal law.

Levinas endorses a similar model of ethical reflection in his remarks on justice. Again, he thinks that there are moments in our lives, face-to-face encounters, when the value of human life imposes itself on us with disarming clarity, but acting on this insight is fraught with complications. We inevitably find ourselves juggling responsibilities and asking whether certain parties have a greater claim on our attention. These complications force us to ask what would be the fairest way to apportion our time, attention, and resources. They impel us, in short, to accept something like the perspective of Kant's categorical imperative: Are we acting in ways which we would have all persons accept?

This is not to say that there is a perfect fit between the two thinkers. Levinas noted in interviews that he felt that there was too much emphasis in Kant on "pure universality."[18] He does not necessarily think it is wrong to equate ethics with universality; he is not interested in, as he puts it, "renewing" ethics with respect to universality. However, he does feel the need to remind Kantians that "ethics arises in relation to the other and not straightaway by a reference to the universality of a law." So, the problem with Kant is not that he recommends bad *procedures* for determining moral conduct. He simply misconstrues their source. Ethical rules need to be rooted in the insight that each life is unconditionally valuable, and it is only on the basis of the challenges involved in respecting this insight that we are driven to develop them.

Levinas might worry, in turn, that there is too much emphasis in Kant on formalism, on rule following for its own sake. For example, Kant argues that we have a duty to tell the truth. He reaches this verdict by showing that we cannot will a maxim that advises deception to become a law of nature without generating contradictions. Levinas might accept Kant's argument as a useful demonstration that it is just or fair to be honest with others, but he would urge us to bear in mind that what licenses this norm is the hope that it helps us fairly balance our responsibilities to others. What matters is not so much the virtue of being honest in itself but the people whom we honor through our honesty. So, one can imagine Levinas parting ways with some of Kant's applications of the rule that we should always tell the truth. He might be more willing to excuse white lies or, to use a notorious foil to Kant's ethics, lies told to protect innocents from persecution on the grounds that our fidelity should ultimately be to the lives serviced by this rule, not the rule itself.

So, Levinas's fit with Kant is not perfect. He is generally sympathetic to Kant's ethics, but he worries that Kantians tend to lose sight of the basic sense of another person's dignity that prompts their tests, and this insensitivity can have practical consequences. Levinas sometimes discusses this point by noting the roles that "charity" and "justice" should play in regulating each other. "Charity is impossible without justice," he remarks, "and justice is warped without charity."[19] On the one hand, a moral theory that roots itself exclusively in feelings of pity or compassion, in charity, is too easily corrupted. Uncritical compassion is too easily exploited and too easily lapses into favoritism. On the other hand, an account of justice that forgets its roots in charity too easily lapses into mechanically enforcing abstract rules at the expense of the lives they are intended to protect.

Thus I am led to return to my earlier conclusions. Levinas seems to be broadly sympathetic to liberal ideals; he accepts the idea that each person has an inalienable dignity that confers on her certain rights.[20] He thinks we should regulate our actions using principles of fairness, principles developed by asking what sorts of rules we can reasonably expect all persons to accept. Finally, he argues for such principles because he thinks they best express our sense that the lives of others are unconditionally precious, and our aim in accepting them should be to ensure the basic well-being of each person. So, it is reasonable to think that Levinas might favor somewhat more aggressive social welfare programs than do other liberals.

4. Making Sense of Levinas's Silence

I want to consider an objection that someone familiar with Levinas will be apt to make and that will help me clarify my proposal. Throughout my book, I have argued against what I deem to be overly mystical, obscurantist interpretations of Levinas. These readings take his reticence as to what ethics he favors to imply that he cannot advance one: he does not defend a particular normative ethics because such theories invariably misrepresent the brush with ineffability that occurs in the face-to-face encounter with the other. I argued against this hypothesis in the preceding chapter, and most recently I claimed that hints of an ethics can be found in him. However, it would be obtuse to deny that these are no more than hints. Levinas simply does not say much about what ethics he prefers. Someone might raise the following objection: What explanation can I give for this silence? If Levinas favors a particular ethical stance, and if nothing prevents him from stating it, why does he not just say so? All things considered, is a more mystical reading not more consistent with this silence?

There are at least three alternative explanations for Levinas's silence. One is fairly straightforward. Clues can be found as to what sort of normative ethics he favors, but his interests are ultimately more metaethical. He is more interested in clarifying the fundamental nature of ethical responsibility than offering advice on how to be ethical. One gets the sense that he thinks we generally have good hunches about what is involved in acting ethically, and our major ethical theories do a fairly good job of reflecting them. We all know it is wrong to manipulate or to harm others; we all know it is good to help others attain happiness. We have trouble, though, reflecting on the ground of these assessments. Levinas looks back on the history of philosophy and sees in it a failure to do justice to the challenges involved in acknowledging others as ethically valuable. His main goal is to correct *this error* by critiquing those trends in philosophy that have hindered our ability to recognize others as ethically valuable and by developing a richer phenomenology of moral experience. In this way, he resembles Kant, whose task in the *Groundwork* is not to offer a new ethics but to help us understand what we unreflectively do in our everyday lives when we make moral judgments, and to shed light on the metaphysical foundations of moral judgments.

Levinas is also reticent, second, because he thinks we tend to misuse our ethical theories. Levinas often cites spontaneous acts of kindness and civility, such as holding open a door for a stranger or greeting a friend, as especially lucid examples of face-to-face encounters, and he often juxtaposes such acts with evaluations one reaches on the basis of more-elaborate ethical theories or political philosophies. Consider, along these lines, one of the passages from Grossman's *Life and Fate* which I discussed in my Prologue, the scene in which a Russian woman shows inexplicable kindness toward a German officer in the aftermath of the siege of Stalingraad:

> The woman could no longer see anything at all except the face of the German with the handkerchief around his mouth. Not understanding what was happening to her, governed by a power she had just now seemed to control, she felt in the pocket of her jacket for a piece of bread that had been given to her the evening before by a soldier. She held it out to the German officer and said: "There, have something to eat."
>
> Afterwards, she was unable to understand what had happened to her, why she had done this.[21]

Levinas cites this scene as exemplifying a face-to-face encounter.[22] Part of what fascinates him is that the woman cannot explain her actions. It is not that her ethical convictions led her to conclude that she should give the officer her bread. Indeed, there is the suggestion that they may have initially prompted her to attempt to kill him.[23] She simply felt impelled to show him kindness at the last moment, without being able to explain why. It is this fact that fascinates Levinas: her empathy for another person extends beyond what her ethical beliefs present to her as reasonable.

Levinas sometimes evinces wariness, then, about ethical theories. There are whiffs of Rousseau in him—the belief that people are basically good, and that ethical and political theories often disfigure this basic goodness. The woman in Grossman's novel shows pity spontaneously, not because she has a thorough grasp of Kant's ethics or Marx's dialectical materialism. Her "little act of goodness" stands in sharp contrast, moreover, to the catastrophic consequences all around her of the clash between two monolithic, totalitarian philosophies—Nazism and Stalinist Marxism. One gets the sense that Levinas fears that our theories often do us more harm than good; we often use them to help suppress our basic impulse to show one another kindness and to legitimate acts of cruelty by justifying them in the name of purportedly higher moral goods.

Factors internal to Levinas's philosophy also make it reasonable for him to question the value of ethical theory. Again, his major complaint with philosophers is that their fixation on metaphysical and epistemological considerations has led them to misrepresent ethical responsibility. In particular, they have tended to delimit the content and scope of ethics based on extramoral considerations. By contrast, Levinas espouses an "ethical metaphysics" in which our sense of the preciousness of human life calls the shots for metaphysics and epistemology. Levinas sometimes indicates that what entices us to reverse this order of priority is that doing so lets us arbitrarily restrict our responsibilities. We have a perverse tendency to seek out metaphysical grounds and epistemological justifications for our ethical convictions not because doing so solidifies our resolve, or sharpens our ability to make moral judgments, but because it lets us shirk our responsibilities and dull ourselves to the needs of others.

Let me give an example. Most ethicists accept the dictum "ought implies can." We are only responsible for actions we can perform or refrain from performing. Imagine a philosopher who cites this idea to argue that it is irrational to feel guilt over the Holocaust. We can be held responsible only for events within the causal orbit of our lives; thus it is irrational to feel guilt for events in the past. Levinas might object, first, that the philosopher is importing ideas from another branch of philosophy, metaphysical views of agency and causation, to do justificatory work in ethics, whereas we should let our sense of ourselves as infinitely responsible call the shots for our metaphysical beliefs. Levinas would also find her argument troubling, second, because it smacks of shirking. As Levinas discusses in "Useless Suffering," the Holocaust poses an ethical crisis for those who live in its aftermath: it forces us to grapple with the question of what constitutes an appropriate *ethical response* to the evil that occurred in it.[24] Is it morally libelous, for example, to go on believing in the existence of an all-powerful, loving God? Or does it instead require some sort of religious response? Levinas would argue that we have a responsibility to make difficult choices about how to live in the face of such questions, and the philosopher's invocation of metaphysical conclusions about causation serves to help us shirk this responsibility.

A lure, then, of delineating ethics on the basis of extramoral considerations is that it lets us restrict the range of responsibilities we recognize as meaningful. Robert Bernasconi may have this point in mind when he suggests that "Levinas suspects that rationality…serves to tame or domesticate morality."[25] To claim that Levinas rejects rationality altogether is, to my mind, too strong, but Bernasconi is right to point out that Levinas worries that our philosophic theories too often serve to "tame" morality's claim on us.

Nor is it only by delineating morality on the basis of extramoral considerations that we tame it; our moral theories can also serve as vehicles through which we restrict the range of responsibilities that we recognize as meaningful. Consider Kant's claim that the respect we feel for one another is ultimately inspired by our rational nature. Levinas would worry that Kant arbitrarily restricts the range of responsibilities we can recognize as meaningful in pinning ethics to metaphysical claims about human nature. If we accept Kant, it will prove difficult, for example, to recognize ourselves as having obligations to nonrational beings, such as mentally disabled persons and the senile elderly. So, one can imagine Levinas having mixed feelings about this aspect of Kant's ethics. He would not necessarily fault Kant for associating humanity with rationality. No doubt Kant is on to something; his ethics sheds light on capacities we need to bear in mind in thinking about how to respect others. However, Levinas would find it dangerous to accept his ethics as supplying us with an exhaustive account of all that is worthy of respect in humanity. For in doing so, we blind ourselves to all that falls outside its purview.

Ethical theories, we might say, are double-edged swords: they help us clarify our responsibilities, but each has moral blind spots. Each sharpens our ability to recognize certain parties as having an ethical claim on us, but at the price of potentially helping us to ignore others. Ethical theories thus come linked with a temptation to misuse them, to use them not to help clarify my obligations to the other but to discredit the other's claim on me.

This brings me to a third explanation for Levinas's silence. My first was that he is more interested in metaethics; my second was that he worries that we tend to misuse our ethical theories. It follows from this last point that our acceptance of any normative ethics will always be somewhat provisional. Consider again my contrast between Levinas and Kant. There is a confidence in Kant that is foreign to Levinas. Kant can feel confident that whenever he encounters an ostensibly rational being, he is required to treat it with respect, and he can feel equally confident that he is not obligated to extend this respect to nonrational beings. However, Levinas cannot draw such clean distinctions. For he would fear that we risk blinding ourselves to humanity in accepting any single account as definitively capturing what is worthy of respect in humanity. Humanity is not, for him, so much something one describes or knows, as something we find ourselves impelled to honor, and our ethical theories are ultimately just tools we use to clarify what is involved in honoring humanity. It follows that our acceptance of any theory will be somewhat provisional. Kant's ethics is no doubt valuable in that it clarifies what is required in order to, say, honor the ability of others to make reasoned decisions about their lives, but our acceptance of it needs to be provisional, just one tool among others that we should use to clarify our responsibilities.

Now, it may seem that I am capitulating to what I earlier stigmatized as a mystical reading of Levinas, one that proposes that he does not advance a normative ethics because no ethics can express the brush with ineffability that occurs in the encounter with the other. I seem to be making the same point: he never offers a normative ethics because none can claim to get humanity right. There are significant differences, however, between my approach and the mystical one I deny. Notice that the mystical reading would have Levinas fault Kant for defending an inaccurate account of human nature, for according to it, humanity is a *something-about-which-nothing-can-be-said*. Levinas is not committed to any such claims about Kant's inaccuracy, on my reading. Indeed, he can praise

Kant's ethics for providing us with a helpful framework for assessing one set of responses we owe to others. The problem is that unthinking, single-minded devotion to this framework is dangerous. Levinas would not have us dismiss existing ethical theories as necessarily misleading; he simply cautions us to remain vigilant in our use of them.

Levinas's reticence, then, is not so much mystic as it is pragmatic. He shares the pragmatic view, espoused by figures such as James and Dewey, that moral theories are tools we use to guide interpersonal action and to resolve the various dilemmas we confront in this sphere. Just as many pragmatists were reluctant, in turn, to stipulate a single, unchanging moral theory on the grounds that there could be no test for one, beyond its success in helping us resolve such crises, so too Levinas is reluctant to set down a single account of justice because the ultimate test for any account is whether it leads us to, in fact, aid the other, and a certain flexibility may be needed to discern which theoretic "tool" best helps us aid her in the face of a given dilemma.

There is a further connection with pragmatism here, as well. Richard Rorty and Hilary Putnam have associated pragmatism with what they call "the rejection of the spectatorial standpoint." I find a similar idea in Levinas. For Levinas, the encounter with the other "de-neutralizes" existence.[26] It renders it morally illicit for us to think of ourselves as passive, neutral observers onto existence, and obligates us to adhere to a standpoint of engaged agency, to view ourselves as, first and foremost, *doers*, beings tasked with *intervening* in the world for sake of the other. Levinas's worry is that we have an all but perverse drive to misrepresent this condition, to seek out opportunities for quietism by lapsing into spectatorial standpoints. Ethical theories constitute one of many crutches we exploit to affect this shift in perspective. Just as there is a temptation to view the other as an object of comprehension, so too there is a temptation to use ethical theories as theoretical surrogates for actually helping the other. Rather than actually *acting* on the other's behalf, we devote attention to debating over whether we have an accurate conception of the good. A final reason for Levinas's silence then is that he fears that outlining an account of justice will feed into his readers' temptation to shirk their obligations. He fears that it will encourage them to make the mistake of assuming that the key issue is whether one understands what is good, just, or fair, and whether Levinas's account of justice accurately conveys these ideals, when the relevant issue is instead whether we are in fact acting to help the other. By not specifying a single account justice, arguably, Levinas may be encouraging his readers to maintain this focus on engaged practice rather than idle, spectatorial theorizing.

Now, I noted at the start of this chapter that my characterization of Levinas as a pragmatist may be seem to be at loggerheads with my earlier portrayal of him as a Kantian liberal. Strictly speaking, there is no necessary contradiction between these two claims. However, the connection between them may seem somewhat opaque, and my proposals in this chapter may seem somewhat fragmentary. Let me end, therefore, by offering an initial sketch of the overall approach to Levinas's ethics implied by my argument in this chapter as well as the previous one. My task in the next two chapters will be to amplify on each of my observations in this chapter—to assess, first, Levinas's compatibility with updated forms of Kantian of the type espoused by John Rawls and Jürgen Habermas, and, second, to explicate his similarities to pragmatism and to show how a pragmatic interpretation of him can develop a positive account of justice. I will have more to say, then, about the claims I have begun to make in this chapter. For now, though, let me offer an initial sketch of the overall approach to Levinas that I plan to defend.

5. A Pragmatic Interpretation of Levinas

The reading of Levinas I have been developing is similar to one that Jean-François Lyotard outlines in *Just Gaming*. Lyotard praises Levinas in *Just Gaming*:

> This is what, for me, makes the thought of a Levinas so important: it shows that the relation with the other, what he calls "the Other" of the "the absolutely Other," is such that the request that is made of me by the other...can never be justified. The model here is the

> relation of God to the Jewish people, with God's initial statement to Moses: "Let them obey me!" At that point it is not known what is to be the content of the act of obedience; only later will it be known. First, there is the prescription to put myself in a prescriptive situation, which is actually something very strange: it is to put oneself in a condition of listening to a discourse that does not describe something but that prescribes an activity, a doing.[27]

For now, I want to pass over Lyotard's claim that Levinas is important because he portrays the other's request as something that can "never be justified." What interests me is his description of the relation to the other as "strange" because it involves a "prescription to put myself in a prescriptive situation." Lyotard reads Levinas as arguing that we discover something intrinsically prescriptive in the encounter with the other—an entity who imposes a prescription to create prescriptions, a moral command to create moral commandments. I made similar claims in the last chapter. Levinas should be read, I argued, as an extremely thoroughgoing moral realist. We uncover an intrinsically normative "moral fact" in the encounter with the other, a moral command on me to establish moral commandments, and he is concerned to draw attention to the ways in which discerning this source of value differs from other phenomena.

Lyotard makes similar observations, moreover, about how the human other resists comprehension. I claimed in the last chapter that Levinas maintains that the other resists comprehension not because it is ineffable but because our relationship to the other is defined in terms of responsibilities. His point is not so much that we cannot comprehend the other, but that it is misleading to depict the relation to her in cognitive terms. The other is not, first and foremost, something I *know* but something I *serve*. Similarly, Lyotard points out that the relation to the other can only be expressed in "a discourse that does not describe something but that prescribes an activity, *a doing*."[28]

Where I differ from Lyotard is in his pessimism about the possibility of developing a positive account of justice. Lyotard points out that Levinas's remarks on the relation to the other undercut a standard argument philosophers have used to formulate and to justify their theories of justice.

Typically, philosophers have defended their ethical theories by seeking to show that they correspond to what is good or just, through what Lyotard calls "mimesis of an essence of justice."[29] Levinas's analysis of the relation to the other undercuts this process because it shows that it is inappropriate to depict this relation as though it were an object of description. Again, the other is not an *object of knowledge* but *a source of responsibility*, something one acknowledges through one's actions, through deeds or "doings." The conclusion Lyotard draws is that it is impossible to justify any account of justice, for doing so invariably relies upon descriptive, mimetic language. Thus one has his praise for Levinas on the grounds that Levinas's shows that the other's request "cannot be justified."

My own view is that Lyotard is wrong to draw this skeptical or pessimistic conclusion. My sense is that he is led to it based on a misunderstanding of Wittgenstein's idea of language games. Consider how he justifies his pessimism about developing a positive account of justice:

> Any discourse meant to account for prescriptions transforms them into conclusion of reasonings, propositions derived from other propositions…In such a derivation…what is derived or deduced is not the prescriptive itself but the citation of the prescriptive, that is, the image, the representation, in the linguistic sense, of the prescriptive…
>
> In the in final analysis, your question supposes that it is possible to derive prescriptive, prescription, obligation in general…To which I say: It cannot be derived. The question you are asking is that of the relation between two language games: that of prescription and that of description.[30]

Lyotard makes it sound as though it is impossible to defend a particular account of justice, or normative ethics, for the same reason it is impossible to simultaneously play gin rummy and poker with the same deck of cards. The rules that define one game are incompatible with the other; similarly, the rules that define language games of prescription are incommensurable with those of description.

However, it is not clear that Wittgenstein believed that language games had such clear-cut rules, or, more generally, that language consists in a grammatical framework of rules, and a more plausible interpretation of him is that we cannot antecedently determine what rules, criteria, or conventions guarantee intelligibility among speakers within a given language game.[31] One simply needs to plunk down and converse with others, sussing out, through these lived conversations, whether we agree about what it makes sense to say within a given language game. Unlike playing Monopoly, where I can show that I am right to collect $200 every time I pass "Go" by consulting the rules on the game's box, there is no way to know what are legitimate "moves" in a language game apart from dialoguing with others and seeing how far we agree with one another. It follows that Lyotard's pessimism is unfounded: for there is no way to antecedently determine what one can and cannot do within the language game of prescription. One simply needs to sit down and to begin talking about what one thinks it makes sense to say about justice.

Now, my point is not to quibble with Lyotard over Wittgenstein interpretation, and I do not want to overstate my criticisms. I agree with the bulk of his claims about Levinas. He is right to point out that the relation to the other resists comprehension, and he is right to link this resistance with how the other renders it morally illicit to adopt neutral, disenageged stances toward her. I do not wish to dismiss Lyotard's insight that we discern a "prescription that there be prescriptions" in the other or that this insight leads Levinas to insist that the other resists conceptual grasp. I also agree with Lyotard that it follows that a certain justification for ethical and political theories is illegitimate. Levinas is troubled by our tendency to distort the relation to the other, to depict it in terms of cognition rather than responsibility, and to exploit this distortion so as to excuse quietism. Ethical theories are one crutch we use to effect this misrepresentation; we use them to delude ourselves into thinking that the task of ethics consists in *knowing* rather than *acting*, in achieving a clearer understanding of the good, of getting our ethical theories to square more accurately with moral facts, rather than recalling that they are no more than tools we use to aid us in actually helping others, and thus the pivotal question is not whether they are *accurate* but whether they are *efficacious*. Lyotard is right, then, to observe that it is a mistake to portray ethics as a matter of achieving a "mimesis of an essence of justice."

My concern is that he too hastily concludes that it is impossible to have *any* discussion about justice. My feeling is that we can discuss whether one account of justice is superior to another, and this discussion need not rely on any objectionably descriptive language. Here it seems to me that we can select between competing accounts of justice based on pragmatic considerations.[32] Levinas depicts justice, again, as a tool we use to negotiate the various challenges we face in balancing our responsibilities to multiple others. One discovers in the encounter with the other that one is infinitely responsible for her well-being, but this realization is always qualified by the awareness that one is equally responsible for all other "others" as well. An account of justice, or a system of moral norms, is simply a tool we use to guide our actions in the face of these complex demands. However, it follows that we can assess our theories of justice the same way we assess the efficaciousness of other theoretical tools. Just as we can ask whether one scientific theory is, on balance, more effective than another, so too we can ask whether an ethical theory, or an account of justice, has merits that render it preferable or superior to others in resolving a set of ethical dilemmas. Ethical theories that, for example, offer proposals for how to convey a greater range of our moral intuitions in our actions, theories that offer more nuanced or more creative proposals, and theories that evince greater clarity or precision in their recommendations seem to be, on balance, superior to those that require us to suppress some of our core moral intuitions and that issue vague or ad hoc recommendations for how to convey them in our actions.

Reading Levinas this way, as upholding a pragmatic view on how to select between competing sets of moral norms, gives me an alternative explanation for why he offers qualified endorsements of Kantianism and Kantian liberalism in his writings. His suggestion is that Kantianism is an effective starting point for assessing how to serve multiple others. It is one of the more effective tools we have developed for modeling how to fairly adjudicate our time and attention. Now, I will leave it to the next chapter to assess the degree to which Levinas endorses this theoretical tool. The point I would make for now is that his grounds for initially recommending it are pragmatic. His suggestion is not that Kantianism is *accurate*, but that it is incredibly *useful*. The further point I would stress, one I develop in more detail in coming chapters, is that it follows that one can generate an account of justice based on Levinas's remarks on the encounter with the other. Lyotard's pessimism about justice is unfounded.[33] For while it may be inappropriate to speculate as to whether an account of justice is accurate, just as it is inappropriate to treat the other as being primarily an object of knowledge, it does not follow that we cannot assess the merits of various accounts of justice in helping us resolve the real-world dilemmas we face in balancing our responsibilities to multiple others.

Endnotes

1. See, for example Rorty, *Contingency, Irony, Solidarity* (Cambridge: Cambridge University Press, 1989), see especially the chapters that comprise Part I. Rorty rejects attempts to justify liberalism on the basis of "transhistorical" facts about human nature. He points out, however, that one can still defend liberalism by arguing *not* that it is *truer* than other ethical-political theories, but that it is *more effective* than them in, say, its openness to public dialogue, its success in achieving uncoerced consensus among citizens who do not share a conception of the good, and its low tolerance for cruelty. My suggestion, ironically, given Rorty's complaints about Levinas, is that Levinas upholds a similar view. Like Rorty, he would have us shift from asking whether a given account of justice is *true* to whether it is *effective* in helping us aid others, and, like Rorty, he favors a broadly liberal, Kantian framework as one of the better tools we have developed for this task.
2. Howard Caygill's *Levinas and the Political* is the most thorough, systematic study of Levinas's political thinking. See also Alford, "Levinas and Political Theory," *Political Theory* 32:2 (2004), 146–171; Critchley, "Five Problems in Levinas's View of Politics"; Hendley, *From Communicative Action to the Face of the Other: Levinas and Habermas on Language, Obligation, and Community* (Lanham, MD: Lexington Books, 2000); Herzog, "Is Liberalism 'All We Need'? Levinas's Politics of Surplus," *Political Theory* 30:2 (2002): 204–227. See Samuel Moyn's *Origins of the Other* (Ithaca: Cornell University Press, 2005), and Michael Morgan's *Discovering Levinas*, 1–38, for discussions of how Levinas's views concerning ethics developed in response to postwar and cold-war politics.
3. Levinas makes this point as well in *Totality and Infinity*. He opens *Totality and Infinity*, for example, by distinguishing between the vantage points of "politics" and "true morality." He goes on to reject political outlooks that are not grounded in the appreciation for the uniqueness and unconditional value of life that emerges in the encounter with the other. Even in earlier works, then, works that precede the distinction between ethics and justice in *Otherwise than Being*, there is already present in Levinas the idea that our social and political institutions should be understood as attempts to manifest respect for the other.
4. Indeed, Levinas notes in several places that the two cannot be distinguished in day-to-day life. His distinction between ethics and justice is, instead, a kind

of heuristic he uses to help his readers recognize how the relation to the other informs the ethical norms as well as the social, legal, and political institutions that structure our day-to-day life. For example, he states in an interview that "[j]ustice itself is born of charity" and that the two "can seem alien when they are presented as successive stages" but that "in reality they are inseparable and simultaneous." See Levinas, *Is it Righteous to Be?*, 168–169.

5. Levinas, *Totality and Infinity*, 52.
6. Ibid., 23.
7. Levinas, "Philosophy, Justice, and Love," 119–120. I am elaborating somewhat on Levinas's comments. Levinas does not use terms such as *humanist Marxism, Western Marxism, Eastern Marxism*, or *orthodox Marxism*. He does indicate, however, that what he finds admirable in Marxism generally is its emphasis on facilitating nonexploitative forms of labor and organizing our resources so as to benefit the least advantaged. He faults the emphasis in certain forms of Marxism on what he calls "conquest," and he worries about the "degeneration" of Marxism into Stalinism.
8. Levinas collected some of these essays in his 1972 book *Humanism of the Other*, trans. Nidra Poller (Urbana: University of Illinois Press, 2003), 9–70 (*Humanisme de l'autre homme*). *Humanism of the Other* contains three essays: "Signification and Sense" (1964), "Humanism and An-archy" (1968), and "Without Identity" (1970). In each of essays Levinas faults such things as structuralist conceptions of language ("Signification and Sense") and structuralist social philosophies ("Humanism and Anarchy").
9. Levinas refers at one point in "Ideology and Idealism" to Deleuze's concept of persons as "desiring machines" in *Anti-Oedipus* (1972). See Levinas, "Ideology and Idealism," in *Of God Who Comes to Mind*, trans. Bettina Bergo (Stanford: Stanford University Press, 1998), 3–14. *See esp.* p. 7.
10. Foucault's *The Order of Things* (*Les mots et les choses*) was published in 1966. An impressive review of it appeared in *L'Express*, the French equivalent of *Time* magazine, in May of 1966. The headline of the book-review section of *L'Express* announced Foucault's book as "THE GREATEST REVOLUTION SINCE EXISTENTIALISM," and the caption to the corresponding photograph of Foucault explained that "Man is a recent invention." The first printing of *The Order of Things* sold out within a week of this review. Subsequent reprintings quickly sold out as well, and Foucault's book spent several months on the best-sellers list. People at the time seem to have seen it as emblematic of the backlash in French thinking against existentialism, and it was read as exemplifying this new Zeitgeist. See James Miller's biography of Foucault, *The Passion of Michel Foucault*

(New York: Doubleday, 1993), 148–159, for information on the reception of *The Order of Things*.

11. I borrow this quotation from Miller's *Passion of Michel Foucault*, 150.
12. See Levinas, "Ideology and Idealism," in *Of God Who Comes to Mind*, p. 10.
13. I quote here from the translation of "Substitution" in *Basic Philosophical Writings*, p. 94. Levinas wrote the essay "Substitution" in 1967. It later became the central chapter of *Otherwise than Being* (1974), his second major work. I quote from the translation in *Basic Philosophical Writings* only because it strikes me as being clearer and more accessible. The corresponding passage can be found in *Otherwise than Being*, 127–128.
14. I use "communitarianism" to refer very generally to ethical theories that take things like cultures, societies, and communities as the primary datum of ethical reflection. It seems to me that Levinas would be skeptical of this general perspective. It does not necessarily follow, though, that he would be hostile to a contemporary communitarian like Charles Taylor. See Morgan, *Discovering Levinas*, 263–268, for discussion of possible similarities between Levinas and Taylor.
15. I borrow this definition of liberalism from Judith Shklar. See Shklar, "The Liberalism of Fear," in *Political Thought and Political Thinkers* (Chicago: University of Chicago Press, 1998), 3.
16. Levinas, *Otherwise than Being*, 56.
17. Levinas, "Philosophy, Justice, and Love," 104.
18. Levinas, "Being-for-the-Other," in *Is it Righteous to be?*, ed. Jill Robbins (Stanford: Stanford University Press, 2001), 114–120. *See esp.* p. 114.
19. Levinas, "Philosophy, Justice, and Love," *Entre Nous*, p. 121.
20. See especially Levinas's "The Rights of the Other Man," and "The Rights of Man and Good Will," in *Entre Nous*, trans. Michael B. Smith and Barbara Harshav (New York: Columbia University Press, 1998), 155–158 for examples of attempts to recuperate a liberal concept of human rights within his philosophy.
21. Grossman, *Life and Fate*, 805–806.
22. See for example Levinas, "Interview with François Poirié," in *Is it Righteous to be?*, ed. Jill Robbins (Stanford: Stanford University Press, 2001), 23–83. *See esp.* p. 81.
23. I am extrapolating a little in my interpretation of this scene. We never see the Russian woman deliberating over whether she should kill the officer. She simply grabs a brick in a fit of rage and, just as spontaneously, throws it aside and offers him her bread instead. I do think, however, that we are supposed to interpret her original decision to kill him as supported by her

ethical convictions. In the scene, the Russian soldiers guarding the officer part when she approaches, and Grossman stresses that they part for her even though they are armed. They seem to consider her morally justified in taking revenge on the officer, given how much she has lost over the course of the war. Afterward, too, she finds no fitting explanation for what she has done. The scene is constructed, then, so that her initial decision to kill him appears as a decision she might reach by reflecting on her ethical convictions. This type of discrepancy between moral judgments that follow from one ideology or another versus what Grossman calls "small acts of kindness" is a recurring theme in *Life and Fate*.

24. See Levinas, "Useless Suffering," in *Entre Nous*, trans. Michael B. Smith and Barbara Harshav (New York: Columbia University Press, 1998), 91–101. *See esp.* pp. 99–100. I discuss Levinas's analysis of the Holocaust as posing a unique ethical challenge for faith in "Responding to Auschwitz: Levinas on Suffering, the Holocaust, and the End of Theodicy" (unpublished essay). Richard Bernstein proposes that Levinas's own philosophy can be understood as a "reply" to the Holocaust in "Evil and the temptation of theodicy," in *Cambridge Companion to Levinas*.
25. Bernasconi, "What is the question to which 'substitution' is the answer?" in *Cambridge Companion to Levinas*, ed. Simon Critchley and Robert Bernasconi (Cambridge: Cambridge University Press, 2002), 234–251. *See esp.* p. 237.
26. Levinas, "Transcendence and Evil," *Collected Philosophical Papers*, trans. Alphonso Lingis (Pittsburgh: Duquesne University Press, 1987), 175–186. *See esp.* p. 182.
27. Lyotard and Thébaud, *Just Gaming*, trans. Wlod Godzich (Minneapolis: University of Minnesota Press, 1985), 22.
28. Ibid.
29. Ibid., 19.
30. Ibid., 45.
31. I refer here to Stanley Cavell's reading of Wittgenstein in *The Claim of Reason* (Oxford: Oxford University Press, 1979). Several of the essays in Alice Crary's and Rupert Read's *The New Wittgenstein* (London: Routledge, 2000) also defend this approach to Wittgenstein. See also Steven Affeldt's "The Ground of Mutuality," *European Journal of Philosophy* 6:1 (1998): 1–31 for a particularly thoughtful discussion of this issue in Wittgenstein and Cavell. Saul Kripke also denies that Wittgenstein views language as consisting in grammatical rules in *Wittgenstein on Rules and Private Language* (Cambridge, MA: Harvard University Press, 1982).

Kripke famously argues that Wittgenstein denies that there are any rules that fix meaning in language. Cavell's reading, which I favor, denies this strong claim, maintaining instead that there is nothing beyond our use of language, our lived conversations with one another, that guarantees such rules.

32. It is worth noting that I oppose what seems to be Simon Critchley's current view on this issue for similar reasons. Critchley discusses what he calls, borrowing from Derrida, the "hiatus" between ethics and justice in "Five Problems in Levinas's View of Politics" and *Infinitely Demanding: Ethics of Commitment, Politics of Resistance*. On the one hand, Levinas thinks the encounter with the other imposes a demand for purportedly universal moral norms, a prescription, as Lyotard puts it, to set down prescriptions. On the other hand, Levinas seems to think that these norms must be invented, Critchley writes, "*ex nihilo*, as it were, and [are] not deduced or read off procedurally from a pregiven moral content." Critchley proposes in these recent publications that Levinas's thinking on this issue should be read as culminating in an "anarchist metapolitics." Now, Critchley is, I find, a little ambiguous on this point. However, he seems to think that there are at least two ways in which Levinas's "hiatus" between ethics and justice suggests anarchism. First, Levinas seems to think that there can be no general procedures for inventing moral norms, simply the decisions we make as individuals. Second, this view leads, Critchley argues, to a form of anarchism that seeks to undermine all efforts to ground moral and political norms in anything other than individuals' decisions. See Critchley, "Five Problems in Levinas's View of Politics," 180, 182–183.

I agree with Critchley that our use of moral norms is, for Levinas, situational and context inflected. I agree because I read him as pointing out that such norms are simply tools we use to address the complex dilemmas we face in acting to help one another. Thus we are continually tasked with "recalibrating" these tools. Where I disagree is with Critchley's and Lyotard's pessimism about the possibility of having any sort of straightforward, positive discussion about our development and use of these norms. Both are guilty of a certain failure of philosophic imagination. Lyotard assumes that *nothing* can be said about why one set of moral norms better expresses the relation to the other than others because this relation does not involve description. Similarly, notice how Critchley equates the impossibility of "deducing" moral norms, or reading them off "procedurally," with inventing them *ex nihilo*. Both conclude that our use of moral norms must be an inexplicable, *ad hoc* process since it cannot be grounded in description or deduction. However, one need not draw this extreme conclusion. For we can discuss whether one set of norms is more effective than another in helping us serve the other. Put another way, one need not lapse into questionable descriptive or deductive language to assess competing moral norms; one can assess them based on pragmatic considerations.

I amplify on this suggestion in chapter 5. For now, I would just add that Critchley's recent anarchistic reading of Levinas leads to a counterintuitive view of how moral norms feature in decision making. It seems odd to say that we invent them *ex nihilo* when we are faced with dilemmas. Suppose I am faced with the dilemma of whether to tell a lie to a friend, one that may spare her pain. It seems strange to say I invented *ex nihilo* a norm such as *One ought to let others make decisions for themselves, even if providing them the information they need may hurt them.* Did I really have no inkling, prior to that moment, that one should respect others' decisions? Did I really invent this norm *out of nothing*? The situation seems to be the following. I was faced with a dilemma, and in evaluating it, I found myself contemplating multiple, preexisting norms, both the one above and a competing norm such as *One should do what one can to alleviate the suffering of others.* I decided that one of these norms would be more relevant than the other in helping me aid the other. Now, my point is not that it is always good to be honest. One could reach the conclusion that it will be better to withhold the information because of the great harm it will do the friend. My point is that *when* we reach such conclusions, it does not seem to be based on some wholly *ad hoc*, unprecedented invention of a new moral norm. It seems to instead involve balancing various broadly pragmatic considerations which are available for discussion and evaluation.

33. This criticism of Lyotard may be inappropriate. I confess that I find his remarks on justice in *Just Gaming* confusing. At times, he makes it sound as though it is impossible to specify a theory of justice for the reason noted earlier—because of the difference between descriptive and prescriptive language games. Lyotard here makes it sound as if the problem is that one cannot speak about justice because one cannot describe it; for as soon as one attempts to describe it, one is no longer playing the language game of prescription. I deny this conclusion for the reasons noted above.

Elsewhere, he gives a different argument. Lyotard observes (p. 45) that we can only create an "image" of a "you must," and he adds that we can state the "content" of a moral command but not "the fact of the 'you must.'" Here, he makes it sound as though the problem is that merely *outlining* a theory of justice does not, of itself, guarantee that people will *act* in the ways it prescribes. My above arguments may not apply to this second argument. However, this second argument is counterintuitive. For one thing, it does not establish that we cannot give positive accounts of justice, only that human beings retain the free will to refuse to act on them. Second, this second argument does not illustrate anything provocative about descriptive or denotative language. The same point can be made about prescriptive language: it too does not, of itself, force people to comply. Merely issuing the command "Clean up your room!" does not, of itself, force people to spontaneously begin tidying up.

Chapter 4

The Duty to Care

Charity is impossible without justice, and justice is warped without charity.

—Levinas, "Philosophy, Justice, and Love"

Although justice and caring have often been seen as distinct, even opposing, virtues, the arguments put forward in this book press for a different view. A justice which does not incorporate the need to respond to vulnerability with care is incomplete, and a social order which ignores care will itself fail to be just.

—Eva Kittay, *Love's Labor*

1. Introduction

Several philosophers have recently argued that Levinas's ethics can and should be integrated into an updated version of Kantian liberalism of the type found in John Rawls's or Jürgen Habermas's writings. For example, Simon Critchley has argued that wedding Levinas to Habermas will lead to a "reciprocal rectification" of each philosopher,[1] and he elsewhere argues that Levinas's "perfectionism" should be supplemented with either Habermas's or Rawls's theory of justice.[2] Steven Hendley similarly argues that Levinas's ethics is compatible with Habermas's and that assimilating the two can correct each philosopher's shortcomings.[3] I noted in the last chapter that there are numerous reasons to think that Levinas would accept some form of liberalism. Yet it is not clear that there is a perfect fit between him and Kantians like Rawls or Habermas. In particular, it is not clear that someone who prizes caregiving as greatly as Levinas can content himself with the way they privilege ideals like fairness, impartiality, and reciprocity.

My aim in this chapter, then, is twofold. I want to refine my claim in chapter 4 that Levinas supports a broadly Kantian ethical perspective by clarifying the extent to which his account of justice can be associated with Kantian liberalism. I plan to do this by debating with philosophers who overstate Levinas's fit with Kantianism. Levinas indicates in several places that he is sympathetic to Kant, but the fit, I shall argue, is not a perfect one.[4] Differences between them become clear, in particular, when we consider the special health needs of dependent persons. Levinas would find Kantian theories of justice too stinting in the care they allot such persons. Identifying Levinas's differences from Kant on this issue also helps me fulfill one of my opening promises. I claimed in chapter 1 that it is unfair to fault Levinas for defending a hopelessly vague or impractical ethics. I directly argued against this criticism in the last chapter by identifying passages where he indicates what sort of normative ethics he favors. However, it also stands to reason that if I can identify modifications he would recommend to Kantian theories of justice, then it must be unfair to fault him for not being able to recommend *any* policies whatsoever. My second goal in this chapter, then, is to make good on my earlier promise to demonstrate that his ideas about ethical responsibility can be put into practice by modeling how they can be used to improve upon Kantian theories of justice.

2. Clarifications and Caveats

There are reasons to think Levinas would defend some form of liberalism. Evidence for this claim was reviewed in the last chapter. Again,

Levinas criticizes antihumanist viewpoints like orthodox Marxism and nationalism, among other perspectives, for privileging abstract social or historical processes over the well-being and rights of individuals. He also indicates in several places that he does not wish to question the idea in liberalism that each person has an inherent dignity; he simply wants to show how this awareness of humanity's dignity arises not out of one's sense of oneself as, say, fundamentally free, but out of the respect for others we most keenly discern in face-to-face encounters with the other. Thus Levinas speaks of his work as culminating in a "humanism of the other." He accepts liberalism's thesis that each person has an inalienable dignity; he simply thinks that it is through the lives of others, the respect they inspire in us, that we first experience humanity as possessing such dignity.

So, I do not deny that it is reasonable to associate Levinas with some form of liberalism. I simply question recent efforts to assimilate him to Kantian liberalism. I use the phrase "Kantian liberalism" to refer to two things. I use it, first, to refer to any liberal theory inspired by Kant. Kant is often associated with the idea that we are essentially autonomous agents; so, a Kantian liberalism might defend a liberal respect for personal freedom by arguing that we are, first and foremost, creatures who can deliberate about what sorts of lives we wish to lead. An interest in fairness and impartiality also sits close to the heart of Kant's ethics; he is interested in identifying rules we can reasonably expect all persons to accept. Thus a Kantian liberal might defend liberal policies on the grounds that they are needed for the open public debate required to generate such rules. I also use "Kantian liberalism" more narrowly to refer to updated forms of Kant's ethics of the type found in Jürgen Habermas's and John Rawls's writings. The term "Kantian liberalism" refers in this narrower sense to recent efforts to refine Kant's ethics by purging it of its reliance on contentious philosophic assumptions, to, as Rawls puts it, defend an ethics that is Kantian "by analogy not identity."

Let me say more about these efforts to salvage Kant. I focus on Rawls, as he will play a more significant role later in this chapter. Rawls admires the egalitarian ideals of Kant's ethics but feels that they need a new justification. Kant and other liberals err in that they pin their defenses of liberalism on one "comprehensive doctrine" or another. For example, they defend liberalism on the basis of substantive *metaphysical* claims about human nature, arguing that liberal policies best reflect our nature as free beings, or they defend it on the basis of substantive *ethical* claims, such as that personal development is the highest good (Mill, Humboldt) or that cruelty is the worst thing we can do (Montaigne, Montesquieu). The problem is that these approaches make our acceptance of liberal policies rely on truth claims which not all parties will accept. Liberalism will strike you as credible only if you believe that we are fundamentally free, rational beings, that personal development is supremely good, or that cruelty is the worst thing we can do. By contrast, Rawls wants to develop a "political liberalism" as opposed to these "comprehensive liberalisms." He hopes to develop a form of liberalism that is "freestanding" in that it relies on no "comprehensive doctrines" at all for its justification.

Rawls invites us to think, in turn, about what he calls "the fact of reasonable pluralism." He invites us to think about modern societies where citizens uphold a "diversity of reasonable comprehensive religious, philosophical, and moral doctrines" yet seek to get along with one another.[5] Do persons living under such conditions commit to any principles simply in being willing to negotiate? Do they tacitly make any assumptions about one another as "conditions of the possibility of a reasonable public basis of justification on fundamental questions?"[6] Suppose, then, that you and I have different values but are willing to look for ways to get along. We make certain assumptions about one another simply in being willing to converse. For example, I assume that you have the ability to develop a coherent worldview; I assume you have the "moral power" to develop a "conception of the good." Yet notice that I thereby accept something like the understanding of personhood found in Kant's End in Itself formulation of the categorical imperative. Kant argues for the End in Itself test by noting that we necessarily think of ourselves as beings who set ends for ourselves. Rawls recuperates this idea on a new footing. Unlike Kant, he does not argue that we *are* autonomous agents. Rather, he notes that we make something like this assumption, necessarily attribute to one another a "moral power" to develop and revise a conception of the good, when we dialogue with each other as citizens.

My more narrow use of the term "Kantian liberalism" refers, then, to attempts to refine Kant's ethics by purging it of its reliance on contentious philosophic claims. Rawls admires the structure of Kant's ethics, and he tries to show how it can be defended without committing oneself to contentious metaphysical or ethical truth claims. Habermas is similar; he tries to show how the basic features of Kant's ethics are corroborated by our ability to use language, and thus one need not uphold a questionable view of human beings as fundamentally rational or autonomous agents to accept it. My question is whether Levinas's ethics is compatible with these refined forms of Kantianism, and my suggestion is that he would have reservations about them. He would find much to praise in both Habermas's and Rawls's approaches, but the fit between his characterization of ethical responsibility and theirs is not perfect.

Here I need to give one last caveat. My strategy for driving a wedge between Levinas and Kantian liberalism will be to show that his ethics cannot be wedded to Rawls's political liberalism. This strategy may seem misguided, as commentators have focused more on his fit with Habermas. Both Simon Critchley and Steven Hendley have published substantial material in which they argue that Levinas can and should be wedded to Habermas,[7] but no substantive research has been done, to my knowledge, on his fit with Rawls. Critchley only suggests in passing that Levinas's "moral perfectionism" needs to be supplemented with a more procedural ethical theory, and he suggests Rawls and Habermas as candidates.[8] So, it may seem as if I am barking up the wrong tree in contrasting Levinas with Rawls; I should focus on his compatibility with Habermas if I want to engage with the secondary literature.[9]

My strategy is justified, though, because Rawls is more heavy handed than Habermas. Rawls and Habermas quarreled with each other over whose theory was more "modest."[10] Rawls held that Habermas's ethics is more ambitious since it relies on the truth of a certain theory of language and thus reduces to yet another "comprehensive liberalism."

Habermas also faults Rawls for being immodest. Here is Rawls's gloss of his critique:

> [Habermas] sees his view as more modest than mine, since it is purportedly a procedural doctrine that leaves questions of substance to be decided by the outcome of actual free discussions engaged in by free and rational, real and live participants—as opposed to the artificial creatures of the original position. He proposes, he says, to limit moral philosophy to the clarification of the moral point of view...and to the analysis of the conditions of rational discourses and negotiation. In contrast, he thinks that my view takes on a more ambitious task, since it hopes to formulate a political conception of justice for the basic structure of a democracy, all of which involves fundamental questions of substantive conceptions, which raises larger questions that only the actual discourse of real participants can decide.[11]

One of Habermas's main goals is to make sure that there are institutions in our communities that guarantee each citizen a say in how the community is run. He is reluctant to predict, however, what decisions citizens who accept these initial conditions will reach about, say, how to allocate their resources. It is up to them to reach an uncoerced consensus on this issue by negotiating. Rawls thinks it is possible to go a step further and to predict what decisions they *should* make. Thus Rawls and Habermas each fault the other for defending immodest accounts of justice. Rawls faults Habermas in that his ethics relies on substantive truth claims about the nature of language, whereas Habermas faults Rawls for defending a heavy-handed theory of justice.

The key point for my purposes is that Rawls is more heavy handed than Habermas: he is willing to make predictions Habermas would find too aggressive. Habermas is content to make sure there are institutions in place that give us an opportunity to engage in dialogue with one another and reach an uncoerced consensus about how to live. Rawls thinks we can go further, via the heuristic of the original position, and identify what decisions citizens should make given a shared commitment to reach fair compromises. Levinas would need to find Habermas unsatisfying, then, if I can show that Rawls's political liberalism is not heavy handed enough for him. My argument, in brief, is this: an ethic that values care-giving as much as Levinas's cannot be satisfied with Rawls's political liberalism because it is too meager, too stinting in the care and support it extends to citizens. Yet Rawls guarantees considerably more support than Habermas. So, Levinas cannot rest satisfied with Habermas either.

3. Levinas on Ethical Asymmetry

Many philosophers have felt that Levinas's philosophy is antithetical to Kant's ethics, and Critchley's and Hendley's efforts to wed the two need to be understood as challenges to this status quo.[12] It will help clarify their proposals if I say something about why so many have thought that Levinas must be at odds with Kant. Levinas says in several places that he admires Kant, but his praise tends to settle on Kant's metaethical claims about the primacy of ethics. He also openly worries about the role concepts like universality and reciprocity have come to play in ethics. Philosophers have generally concluded that some critique of Kant's ethics must be implicit in him. He can be read, I find, in two ways on this issue, as committed to either a weak or a strong critique. Let me review his general claims on the particularity and asymmetry of ethical responsibility before explaining each of these interpretations.

There has been a tendency to overstate Levinas's difference from Kant; so, let me note a few similarities first. Both Kant and Levinas are interested in the awe that humanity inspires in us. Kant writes about the respect we feel before others, the way the "mind bows" before the "humble plain man."[13] Levinas is similarly interested in the respect others instill in us when, in the face of their vulnerability, we feel pity for them. Both also describe this respect in similar terms. Kant associates ethics with recognizing unconditional side constraints on our lives, and he presents ethical responsibility as most clearly discerned in cases where it trumps our self-interests. Levinas uses similar language to discuss face-to-face encounters. I experience the world as a place of "enjoyment," and when I have a face-to-face encounter, I feel obligated to forgo things

I enjoy so that I can attend to the authority I find reflected in the other, to, as Levinas puts it, "take the bread out of one's own mouth, to nourish the hunger of the other with one's own fasting."[14]

So, both Kant and Levinas associate ethical responsibility with recognizing in others a source of unconditional value. Levinas would, however, fault Kant's analysis of this value. Kant equates the unconditional character of morality with universality. To say an act is *unconditionally* right is, he reasons, to say it is right *in all circumstances*, hence that it embodies a *universal* law. Levinas thinks this misrepresents our experience of the unconditional character of morality. Suppose I am shamed by images on the news of men and women being victimized; scrutiny of this reaction shows that it is better described in terms of particularity than universality. On the one hand, I feel impelled to help *particular individuals*: I am outraged not because a universally acceptable rule has been flouted but because particular persons are being hurt. On the other hand, I also feel *uniquely complicit* in their victimization. After all, it is up to me to do something in *my life* to address it. Indeed, Levinas thinks our sense of ourselves as unique individuals flows out of this challenge of finding ourselves obligated to enact a fitting response to the needs of others. "Uniqueness seems to me to assume meaning," he writes, "in terms of the irreplaceablility that comes to…the I in the concreteness of a responsibility for the other."[15] Catherine Chalier nicely summarizes these two ways in which ethical responsibility is fundamentally particular: "The [other's] face compels the subject looking at it, to ask of itself: what will happen to that singular other, vulnerable and threatened with violence, if I—also a singular subject—fail to act?"[16]

Levinas would fault Kant's description of ethical responsibility for a second reason as well. Kant thinks that an ethical law is a universal one, and he recommends that we use the concepts of symmetry and reciprocity to test if we may be acting on a universal law. Do I treat others in ways I myself would like to be treated? If not, I have reason to doubt that I am acting on the basis of a universal law. Kant depicts ethics, then, as comprising reciprocal obligations between equals. By contrast, Levinas thinks that ethical responsibility is "asymmetric." Suppose, again, that

I see persons being victimized on the news and feel a deep sense of pity for them. I feel, if only for a moment, that they have an unconditional claim on me. Levinas is fascinated by the purity of our altruism in such moments. For that moment at least, their claim is absolute. The gamut of things I esteem or desire strikes me as secondary to a higher authority: the well-being of others. It is misleading to equate responsibility with reciprocity, then, because the pity I feel for others is remarkable precisely because it involves caring for something I perceive to be *greater* than myself. Levinas makes this point about the asymmetry of responsibility in an often quoted summary of his philosophy: "[W]ith the appearance of the human—and this is my entire philosophy—there is something more important to me than my life, and that is the life of the other."[17]

4. Care Versus Justice: Weak and Strong Readings

Levinas seems to be doubly opposed to Kant's description of ethical responsibility. Kant equates responsibility with *universality*, whereas Levinas thinks it should be described in terms of *particularity*. Kant also thinks the best way to determine if an action evinces a universal law is to ask if *equally* placed individuals would accept it, whereas Levinas thinks we experience others as *more valuable* than ourselves in lucid moments of moral perception. There are two ways to interpret how he might develop these differences into a full-blown critique of Kant. Let me focus first on the weak critique. I call this criticism "weak" not because it is less convincing or less critical of Kant but because it is less extreme in the view it ascribes to Levinas. On this reading, Levinas thinks there are preferred descriptions of ethical responsibility. In particular, ethics is better characterized in terms of caring than justice.

The weak reading attributes to Levinas a view like that of some "ethics of care" feminists and virtue ethicists who claim there has been too much focus in ethics on rules. Carol Gilligan's *In a Different Voice* is the most famous example of this view.[18] Gilligan argues in *In a Different Voice* that Lawrence Kohlberg's famous studies of developmental moral psychology were biased against women. Women tended to score lower on his tests, but Gilligan notes that empathic identification plays a larger role in their deliberations, and their scores improve if we allow a second form of ethical deliberation that views ethics not as a matter of applying rules but acting on the basis of loving concern for particular individuals. Other feminists have expanded this criticism, arguing that a care-based approach is not only a legitimate alternative to more rule-focused approaches but better expresses what ethics is all about. Philosophers associated with virtue ethics have made similar arguments. For example, Michael Stocker faults Kant's ethics in "The Schizophrenia of Modern Ethical Theories" for being myopically focused on duty, when duty and impartial ethical rules represent "only a small part, a dry and minimal part, of ethics."[19]

The weak reading of Levinas attributes this type of critique to him. He claims that responsibility is better described in terms of particularity than universality: ethics is more a matter of acting on the basis of a loving concern for particular individuals, and Levinas seems to fault impartialist approaches in ethics because they distort this primordial sense of responsibility by equating ethics with rules. Similarly, care-based feminists argue that ethics involves empathically identifying with "concrete others" and fault impartialist theories for distracting us from this fact. The weak reading interprets Levinas, then, as arguing that there are preferred descriptions of ethical responsibility. In particular, ethics has more to do with care than justice, with acting on the basis of loving concern for individuals, and cultivating loving relationships, than rigidly adhering to impartial rules of conduct.

Let me shift to the strong reading, which maintains, against the weak reading, that there can be no preferred descriptions of ethical responsibility. It helps here to review Levinas's distinction between ethics and justice. Levinas reserves the term "ethics" for the awareness of others as inviolable that occurs in face-to-face encounters; he uses "justice" to refer to the systems of moral norms we develop based on this awareness. The strong reading stresses the tension between ethics and justice. Levinas thinks there is something literally miraculous about our awareness of others in lucid moments of pity and compassion. Yet he also seems to think we are obligated to subject this awareness of others as transcendent to exposition insofar as we need to develop moral norms that reflect it. How can he make these two claims? He seems to be saying both that we *cannot* put our experience of others into words and that we *must* find words for it.

The strong reading of Levinas points out that he seems to see justice as a necessary evil. We cannot get along without systems of moral rules, but these rules cannot help but misrepresent the experience of ethics that prompts them. Thus Levinas sometimes states that every theory of justice invariably requires violence.[20] The strong reading stresses this idea that there can be no nonmisleading, nonreductive, nonviolent descriptions of ethics. The problem with Kant, in turn, is that he thinks it is possible to describe ethical responsibility. In particular, he errs in assuming it can be paraphrased in terms of universality. Unlike the weak reading, though, the strong reading finds it no more accurate to paraphrase one's sense of others as unconditionally valuable in terms of care or love. For that too would err in assuming that one can accurately describe the apprehension of humanity that emerges in the face-to-face encounter with the other.

5. Marrying Levinas and Kant

I noted earlier that some philosophers have argued that Levinas's ethics can and should be wedded to a Kantian theory of justice. Steven Hendley and Simon Critchley have offered the most compelling arguments for this proposal.[21] Each can be read as replying, respectively, to my weak and strong readings. Hendley takes Levinas to be advancing a care-based ethical perspective and tries to show that it is consistent with Habermas's discourse ethics. Critchley reads him as positing an ethical foundation for justice that cannot be recuperated in it, and he tries to show that this view is unsatisfying on its own and needs to be supplemented with Habermas's or Rawls's theory of justice. Let me review each of these proposals.

I begin with Hendley. Hendley has two aims in his work on Levinas and Habermas. He wants to show that their views are compatible, and he wants to show that each can refine the other's ethics. Each has nothing to lose and much to gain from being wedded. He focuses on two areas in Habermas to argue for his compatibility with Levinas. He argues that Habermas and Levinas have similar philosophies of language, and he argues that they have similar hunches about the need for impartiality in ethics. I want to pass over his arguments for the first point, as Habermas's and Levinas's philosophies of language are extremely complex, and a comparison would require a chapter unto itself. More crucial for my purposes is his claim that Levinas shares Habermas's intuitions about the need for impartiality.

It is helpful to read Habermas as heir to a debate over Kant's ethics that dates back to Friedrich Schiller. Schiller famously faulted Kant for focusing too much on duty at the expense of feelings of affection like love and friendship. "I must learn to hate my friends," he jibed, "so that my doing them good, which I now gladly do, will acquire moral worth."[22] Kant replied by noting that nothing in his ethics requires him to dismiss feelings like love or friendship as inimical to ethics; acts inspired by them simply need to be regulated by an impartial sense of duty. It is good for a parent to love her child, but it is not good if, based on this love, she threatens a teacher who has caught her child plagiarizing. So there is nothing wrong with being moved to perform moral acts by love, friendship, or compassion. When we reflect on what acts to perform, however, our deliberations should be regulated by an impartial perspective. We should ask ourselves what, ultimately, is the fair thing to do. Defenders of Habermas have made similar arguments in response to critiques by ethics-of-care feminists. Seyla Benhabib uses the example of the care a mafia family shows its members to defend Habermas.[23] It is good to feel a sense of fraternity toward one's friends and neighbors, but not if it leads to *Sopranos*-like violence.

Hendley finds evidence in Levinas's distinction between ethics and justice that he recognizes this need for impartiality. Levinas thinks that the realization that I am "infinitely responsible" for the other is always qualified by the realization that all others have an equal claim on me, and thus there is a need for impartial rules of justice that help me adjudicate competing claims on my attention. Hendley reads Levinas as here departing from the view defended by ethics-of-care feminists. Levinas does not posit two perspectives for ethical deliberation, care and justice, but a *single standpoint* in that care drives us to accept an impartial deliberative standpoint: "[I]t is misleading, I believe, to posit two, distinct moral points of view. For, as we have seen, it is our caring relation to the concrete other that establishes the point of the impartial, procedurally governed perspective of justice."[24] Levinas does not share the partialist convictions, then, of an ethics-of-care feminist. An ethics-of-care feminist would have us focus on the love we feel for those closest to us, but Levinas thinks our care for others requires us to regulate our conduct using impartial rules. Indeed, he makes this point in his essay "Uniqueness":

> The love of one's fellowman, and his original right, as unique and incomparable, for which I am answerable, *tend of their own accord to make appeal to a Reason capable of comparing incomparables, a wisdom of love*...Here, the right of the unique, the original right of man, calls for judgment and, hence, objectivity, objectification, thematization, synthesis.[25]

Levinas seems to be sympathetic to the Kantian ethical procedures that Habermas defends in his work. Levinas would accept the ethics-of-care feminist's insight that ethics begins with a sense of loving concern for individuals, but he thinks that the finest gift we can give in the name of this love is to treat each other fairly. The best way to "love one's fellowman" is to "love wisely."

Simon Critchley also thinks that wedding Levinas to Habermas will lead to a "reciprocal rectification" of each philosopher. He begins, however, with something like my strong reading. Crtichley was one of the first philosophers to challenge the popular view of deconstruction as a nihilistic, apolitical philosophy, and to argue that it is rooted in ethical ideals. He defends this alternative view of deconstruction in, among other places, a series of exchanges with Richard Rorty.[26] Rorty characterizes

Levinas and Derrida as "ironists" whose work has no political significance. Critchley notes that both distinguish ethics from justice: both argue that moral and political systems flow out of an experience of ethics they cannot recuperate in their strictures. Critchley argues that this perspective lends itself to a form of political critique, and thus Rorty's portrayal of Levinas as an ironist misses the mark. For, given any account of justice, we can feel confident that it misrepresents in some way its underlying foundation in ethics. We may not be able positively to state what it means to recognize others as unconditionally valuable, but we can explore how accounts of justice invariably misrepresent this point.

Critchley begins, then, with something like my strong reading. He argues that it is misleading to fault Levinas's philosophy as having no political import, but reaches this conclusion by reading Levinas as claiming that there can be no nonmisleading accounts of justice. Critchley questions the adequacy of such an outlook in his more recent writings. Levinas may have useful things to say about how justice can be "deconstructed" into ethics, but we live in a world in which we are faced with urgent yet complex moral dilemmas. We want advice not only on how to *critique* existing moral theories or political systems but how to create *more humane* ones in their place. It does not look as if Levinas can offer any advice, for it is his belief that any claim to have captured the preciousness of humanity in a theory of justice will be misleading. So, it looks as if we have, as it were, a moral obligation to develop a procedurally robust normative ethics yet cannot generate one based on Levinas's characterization of ethics. Thus Critchley refers to the gap between ethics and justice as "the fundamental *aporia* of deconstruction."[27]

It is for this reason that Levinas needs someone like Habermas. Levinas admits the need for procedures for determining moral conduct, but his analysis of ethics prevents him from advancing any. Habermas's ethics is rich with such tests. Marrying Levinas to Habermas, then, can correct this flaw. Critchley has made a similar argument by positing two species of moral philosophers. He cites Rawls and Habermas as examples of "moral legislators," ethicists who develop "detailed precepts, rules and principles that add up to a theory of justice," whereas he refers to Levinas as a "moral perfectionist" who thinks that "ethics has to be based on some form of basic existential commitment or demand that goes beyond the theoretical strictures of any account of justice or socially instituted ethical code."[28] Critchley proposes a division of labor in ethics: "Surely the entire difficulty of moral theory and moral life consists in the fact that we require *both* legislators *and* moral perfectionists, both a compelling description of ethical demand and a plausible theory of justification for moral norms." We can turn to Levinas when we need stirring reminders of the urgency of ethics and of the enormity of the demands it imposes, and we can turn to figures like Rawls and Habermas when we require detailed advice on how to fulfill them.[29]

6. Forced Marriages: Questioning Levinas's Compatibility with Kant

Hendley and Critchley reach similar conclusions, despite their different readings of Levinas. Both argue that Levinas can and should be wedded to some form of revised Kantianism and Kantian liberalism. To some extent I agree with their claims. As I argued in the last chapter, there is evidence that Levinas is sympathetic to Kant's ethics. However, the fit is not perfect. Levinas esteems caregiving significantly more than Kantians like Rawls and Habermas, and I will argue that substantial differences emerge when we consider their stances toward issues like the rights of disabled or dependent persons to exceptional forms of health care. Even apart from this criticism, though, it seems to me that Critchley's and Hendley's proposals are flawed.

Let me focus on Critchley.[30] Critchley argues that Levinas's work leads to paralysis in ethical deliberation and proposes that he should be supplemented with Habermas or Rawls. However, suppose we ask this question: *Why Kantians*? Why should we think Levinas would feel uniquely drawn to Habermas or Rawls? Critchley makes this marriage sound arbitrary. Consider how he defends his proposal in response to an essay by Axel Honneth:

> I think Honneth's paper opens the possibility of a reciprocal rectification of the two philosophic currents that could be said to define the conflictual space of European philosophy, namely the Frankfurt School of Critical Theory on the one hand, represented by Habermas, and 'postmodernism' or deconstruction on the other, represented by [Levinas and] Derrida.[31]

It sounds as if Levinas should wed Habermas only because there are no other suitors. There are only two options in European philosophy: deconstruction or Critical Theory. We have a responsibility to develop a robust theory of justice, and Levinas's quasi deconstructionism is unhelpful in generating one. So, why not give the other option a shot?

Critchley can perhaps object that there are reasons to favor Habermas over other suitors. As I noted in the last chapter, Levinas suggests that his thinking is similar to Kant's. Critchley could add that Habermas's brand of Kantianism is more sensitive to the contextual nature of moral deliberation and was developed with an eye toward meeting the criticisms of postmodern philosophers. So, it is not just that Habermas is Levinas's *only* suitor in continental philosophy: Levinas expressly says he *likes* Kant, and Habermas's brand of Kantianism is the best around.

Yet why should we uncritically accept Levinas's praise for Kant, especially since it centers more on Kant's metaethics than on his normative ethics? Levinas speaks favorably of the End in Itself test in one interview,[32] but most of his praise focuses on Kant's claims about the primacy of practical reason. He also often sounds more like Peter Singer. His paradigm cases of moral actions do not involve activities such as truth telling, but issues of basic survival and physical well-being, "taking the bread out of one's own mouth to nourish the hunger of the other." So, why not attribute to him the type of imperative to redistribute our collective resources so as to address poverty that Singer defends in "Famine, Affluence, and Morality?"[33] Or why not read him as a casuist, as identifying a set of paradigmatic ethical cases in his remarks on face-to-face encounters that can serve as a basis for parallel deliberations in more complex cases? Or, given his contempt for cruelty, why not link him less to liberalisms in the tradition of Kant and more to what Judith Shklar calls "liberalisms of fear," liberalisms that see cruelty as "the worst thing we can do" and defend liberal policies as being maximally resistant to the threat of physical brutality and humiliation?

Let me be clear. I do not think any of these proposals *refutes* Critchley. Indeed, I agree with the basic suggestion that Levinas would welcome Habermas or Rawls. Yet their *prima facie* plausibility points to a gap in Critchley's argument. Critchley's argument turns on something like the ineffability hypothesis; he argues that Levinas *does not* advance an account of justice because he *cannot* do so given his claims about the other's inexpressibility. I am inclined to question Critchley's argument because I deny this hypothesis, but it also seems to me that his commitment to it, or something like it, sabotages his efforts to wed Levinas to Rawls or Habermas in that it prohibits him from dissuading us from wedding Levinas to one of the other figures noted above.

For how can he convince us that Levinas would find Rawls or Habermas more congenial? On the one hand, he can grant that there are better and worse ways to describe the responsibility one apprehends in face-to-face encounters, and evidence can be found to show that Levinas favors a Kantian description over, say, a consequentialist or care-based one. Yet if he makes this argument, it does not seem appropriate to read Levinas as positing an ethical foundation for justice that cannot be expressed in its norms, for he would need to grant that some norms more fully express this foundation. On the other hand, he can claim that there can be no preferred descriptions of ethics, but then it seems difficult to claim that Levinas bears any special affinity to Rawls or Habermas, and his choice of these figures is apt to seem arbitrary.

A further criticism can be made here of both Critchley and Hendley. Both want to pair Levinas up with contemporary Kantians on the grounds that he does not offer enough advice on how to put his ideas into practice. When faced with a thorny moral dilemma, we can then ask how Rawls or Habermas would solve it and attribute their answers to Levinas.

The problem is that simply grafting Kantian tests onto Levinas's philosophy leaves Levinas's conception of justice unexplained. Critchley makes the following observation in arguing for wedding Levinas to Habermas: "[W]hat is lacking in Derrida's and Levinas's work and more widely in 'postmodern' thinking is a full theorization of the passage from care to justice. That is to say, an account is needed for the passage from the quasi-phenomenology of ethical asymmetry to [a] full-blown theory of justice."[34]

Critchley's claim about what needs to be clarified in Levinas is correct: we need to clarify the passage from care to justice. Simply grafting Habermas's or Rawls's tests onto him, however, does not clarify this gap: it leaves it equally untheorized. For Habermas and Rawls simply are not interested in asking how to embody *care* in their principles of justice; they are interested in identifying principles that are maximally *fair*—principles any interlocutor (Habermas) or anyone willing to compromise (Rawls) should find acceptable. However, *this goal* is not Levinas's: his problem is that of clarifying what care writ large in impartial ethical rules might look like.

Along these lines, Levinas remarks in *Otherwise than Being* that "justice remains justice only in a society where there is no distinction between those close and those far off, but in which there also remains the impossibility of passing by the closest."[35] There is supposed to be play between care and justice. One can imagine Levinas faulting ethics-of-care feminists for overlooking the need to temper the immediacy of care with fair judgments; their view too easily lapses into favoritism, into unjustly ignoring those "far off," and showering attention on one's immediate community. However, he would also worry about theories of justice that are not rooted in care, that allow us to "pass by" those who need help in the name of abstract, impartial rules. One gets the sense that Critchley's and Hendley's proposals serve to postpone a difficult discussion. If we want to understand Levinas, we need to discuss what it means to record care's ethical significance in impartial principles. Grafting Habermas's or Rawls's tests onto him does not help us. For we still do not know how a philosopher who prioritizes care will reason in developing such principles, only how two philosophers interested in fairness teased out *its* implications. So, what happens when we have this discussion? Does Levinas still end up defending Kantianism?

7. Care, Fairness, and Dependency Critique

The inability of Kantian liberalism to do justice to care's ethical significance becomes clear when we consider critiques of Rawls by philosophers who expressly are concerned to stress the need for social institutions which are grounded in caregiving. I have in mind critiques by feminist philosophers and philosophers working in disability studies, what Eva Kittay calls "dependency critique." These criticisms allege that Rawls's theory of justice systematically ignores the special needs of dependent persons and places inordinate burdens on caregivers. My suggestion is that Levinas would sympathize with this critique, and Kittay's proposals illustrate better the kind of reasoning he would use in thinking about issues of justice than do Rawls's or Habermas's. In short, he would be drawn to this literature as an example of how the ethical encounter with the other can generate a politics. Let me elaborate, then, Kittay's dependency critique. I will discuss how her critique clarifies the passage from care to justice in the final section of this chapter.

To explain Kittay's critique, I need to explain how Rawls uses his idea of citizenship to develop an "index of primary social goods." Rawls opens *Political Liberalism* by inquiring about the conditions of the very possibility for compromise: what assumptions do we make when we dialogue as citizens who may disagree in our worldviews yet hope to get along? He notes that we attribute to one another two "moral powers." We each assume that the other can formulate a life plan, and we assume that we can recognize fair compromises when they are offered to us. I noted earlier how the former "moral power" lets Rawls recuperate something like Kant's End in Itself test. Rawls thinks that further conclusions can be mined from these two moral powers, for we seem to tacitly accept whatever institutions are needed to cultivate them. We may disagree in our worldviews, but we seem to agree, at least, that we have the ability to construct them, and thus we both should be willing to accept whatever institutions are needed to develop this ability.

Thus Rawls thinks we can develop a list of "basic social goods" implied by the abilities we attribute to one another as citizens. He posits five "primary goods":

1. The basic rights and liberties. (Freedom of thought, liberty of conscience, etc.)
2. Freedom of movement and free choice of occupation against a background of diverse opportunities.
3. Powers and prerogatives of offices and positions of responsibility in the political and economic institutions of the basic structure.
4. Income and wealth.
5. The social bases of self-respect.[36]

I want to bracket the question of how Rawls arrives at this list. More important is how he uses it to specify institutions to which all citizens are entitled. Consider this list's implications, then, for education. We ascribe to one another the moral power of being able to develop and revise a life plan. A condition for realizing this power is that we have a "free choice of occupation against a background of diverse opportunities." Thus all citizens are entitled to a free education, for they must be aware of the range of life opportunities available to them and have the requisite skills needed to pursue their choices.

This example may still sound vague, but consider something as mundane as the fact that children are entitled to transportation to their schools. Rawls would see this practice as mandated by his list. Indeed, he might cite something like the Individuals with Disabilities Education Act (IDEA) as an example. IDEA requires that disabled students be given free transportation to their schools. Schools cannot beg off this obligation by claiming that making their buses accessible to disabled students would require a major financial burden. They cannot, because to do so would exclude certain citizens—disabled persons lacking the financial means to secure alternative forms of transportation—from an institution they need to cultivate the moral powers we necessarily take one another to possess. Rawls's list gives a set of ideals, then, that we can use to specify goods to which all citizens have a nonnegotiable right.

Kittay criticizes Rawls's list of basic goods. "What I deem dependency critique," she explains, "calls attention to the neglect of dependency and the consequences of that omission in theories of equality and social justice."[37] She invites us, then, to consider dependent individuals: mentally and physically disabled persons, young children, and the frail elderly. What distinguishes these persons is that their incapacitation makes them depend on others for their well-being. They are unable function in certain environments, and they must entrust aspects of their lives to caregivers. She alleges that an approach to justice like Rawls's—one that prioritizes fairness—systematically ignores the special needs of dependents. In focusing on fairness, a Kantian account of justice "fails to sustain the egalitarian vision that purports to inform it."[38]

Why should we think that Rawls's account of justice ignores the needs of dependents? After all, I just gave an example in which his list of basic goods was used to justify an institution that benefits disabled students. The problem is not that his liberalism does not have *implications* for dependents, but that it cannot recognize them as having genuinely *unique* needs that entitle them to receive care beyond that guaranteed to other citizens.

Rawls thinks it is possible to develop an approach to health care based on his list of basic social goods. Early in *Political Liberalism*, he asserts that his list covers "what we may call normal health care."[39] He reasons that a base level of "normal human functioning" is required to take advantage of the opportunities protected by his list. He also observes that it is realistic to think that we will all, at times, fall ill or sustain injuries that keep us from maintaining this level of functioning, and so there will inevitably be times when we need further forms of support. Thus Rawls thinks he can argue for a national health care program on the grounds that all citizens should have access to the support they need to sustain a normal level of human functioning. Notice, though, that his aim is to identify a minimal level of support that we should all, in principle, find acceptable given that we would want it for ourselves in case we become ill or injured. More extreme health care needs are not covered by his list. Indeed, I take it that he uses the loaded phrases "normal health care" and "normal human functioning" for this reason.

This point can be seen as well in Rawls's suggestion that discussion of how to handle more extreme health care needs should be postponed until the "legislative stage":

> In [cases of extreme variations in physical capacities], the variations that put some citizens below the line as a result of illness and accident…can be dealt with, I believe, at the legislative stage when the prevalence and kinds of these misfortunes are known and the costs of treating them can be ascertained and balanced along with total government expenditure. The aim is to restore people by health care so that once again they are fully cooperating members of society.[40]

Rawls's point in this passage will make more sense if we recall my IDEA example. I noted that Rawls would see IDEA's stipulation that disabled students receive free transportation to their schools as an example of how his list can be used to identify institutions to which citizens have a nonnegotiable right. Notice that a parent cannot use his list to demand that creationism be taught in her child's school. His list requires that all citizens be given equal access to a basic education; it does not permit a parent to dictate the content of this education. If a parent wants her child to learn creationism, she needs to do it through a private association, such as a Sunday school.

Now, citizens have a nonnegotiable right to health care covered by Rawls's list of basic goods. For example, an African American woman can invoke his list if she is forced to wait in an emergency room while whites receive treatment. Her claim is akin to the disabled child's in my IDEA example: she has a nonnegotiable right to a certain level of support, and she is not being given equal access to it. However, health care needs that fall outside this specification are on the same footing, more or less, as the creationism classes in my second example. For example, a parent raising a child with Down's Syndrome can appeal to Rawls's list for the same level of support it extends to parents of nondisabled children, but she would not be immediately entitled to further aid, even though the complications in her child's birth make it likely that her child will have additional health care needs.

Thus one has Rawls's claim that consideration of how to handle more extreme health care should be postponed until "the prevalence and kinds of these misfortunes are known and the costs of treating them can be ascertained and balanced along with total government expenditure."[41] To some extent, my comparison with Sunday school classes is unfair; Rawls concedes that we should *try* to help persons with more extreme health needs once we know what resources are available to us. Still, his point in postponing this discussion until the "legislative stage" is to underscore that citizens do not have a nonnegotiable right to this additional support. The state may grant it once it knows how much it will cost and what resources are available, but this aid will be supererogatory, a gesture of generosity on the state's part. The problem, of course, is that dependents *do* have special needs. Their incapacitation puts them in a position where they may require long-term, costly care. The upshot of Rawls's liberalism is that they do not have any special moral claim on such care.

Another problem is that Rawls's list is derived from his notion of citizenship and the moral powers we accord one another as citizens. It is not clear that all persons, particularly those with mental disabilities, fall under this ideal. For example, persons with severe forms of mental retardation or persons in the later stages of Alzheimer's may not have the moral powers Rawls equates with citizenship. Their special needs seem to be, in turn, morally invisible. Contrast my earlier example of the student with a physical disability with a child with a mental disability. The child with the physical disability can demand that her school system provide her with transportation to her school, even if doing so requires it to restructure its budget. However, the parents of a child with severe mental retardation cannot similarly demand that the state subsidize the cost of paying a caregiver to take the child for walks while they work. In the first case, the physically disabled child has a right to her education because it is consistent with her moral powers as a citizen, but in the second case the parents' plea on behalf of improving their child's quality of life is morally invisible because the child cannot cultivate these moral powers.

There is a further element to this critique. I have focused on Rawls's denial that dependents have a special claim on care beyond that guaranteed to other citizens. There is also a feminist component to this critique. Additional care beyond that secured by Rawls's list of primary goods is supererogatory: the state may offer it, but has no obligation to do so, and people will typically need to finance it on their own. Proponents of dependency critique are irked that Rawls ignores the special needs of dependents, but they are also troubled because one segment of society, women, has typically had to pick up the slack. Women have typically been assigned caregiving roles as part of their familial duties, and they hold the bulk of the low-paying jobs in day-to-day health care.[42] Rawls's liberalism places oppressive burdens, then, on persons who become caregivers. He would presumably find it unjust for an employer to neglect to pay an employee for her labor. Yet a parent who opts to become a caregiver seems to accept a form of labor for which she can expect no compensation. At the same time, it is not clear that this labor is superfluous. We spend considerable portions of our lives in situations of dependency, not simply, as Rawls suggests, during momentary bouts of illness or injury, but for considerable portions of our childhood and, to a growing extent, the final years of our lives. In neglecting the special needs of dependents, then, his liberalism unevenly distributes an indispensable form of labor and depicts it in such a way that those who accept responsibility for it can typically expect no financial compensation.

8. The Duty to Care in Levinas

My suggestion is that Levinas would admire Kittay's critique as an example of how to generate an account of justice based on the ethical encounter with the other. I argued at the close of chapter 3 that Levinas upholds a broadly pragmatic approach in ethics. He thinks that the human other resists comprehension not because it is wholly ineffable but because the relation to the other is intrinsically normative—the other is not something I claim to *comprehend* but something that *impels* me to serve her through *acts* of charity and justice. Systems of moral norms need to be understood, in turn, as *tools* we use to assist us in this task—tools we use to negotiate the challenges involved in balancing responsibilities—and thus the relevant question with respect to normative ethical theories is not whether they are *accurate*, corresponding to some platonic form of the good, but whether they are *effective* in helping us resolve the crises that beset us in our efforts to serve one another. Moreover, we can discuss whether one set of norms does a better job of aiding us than another, and in this sense, it is possible to develop a positive account of justice based on the ethical encounter of the other. Levinas would simply have us frame this discussion, as do pragmatists, in terms of *efficaciousness* rather than *accuracy*.

Now, my full defense of this pragmatic approach will need to wait until the next chapter. There I discuss in detail how a pragmatic reading of Levinas can develop a positive account of justice by supplying us with procedures for selecting between competing sets of moral norms. It is possible, however, to indicate the drift of this argument, and doing so will help me explain why Levinas would accept Kittay's dependency critique as illustrating how to generate justice from ethics. For the pragmatist, the considerations to be used in selecting between moral theories are not all that different, ultimately, from those used to select between scientific theories. The relevant issue for both is not whether our theories are accurate but whether they are effective in resolving dilemmas. The difference is that we use ethical theories to resolve a different set of dilemmas, those that emerge with respect to interpersonal relations.

Some of the hallmarks of a good scientific theory apply, then, to ethical theories as well. For example, a scientific theory that supplies us with plausible explanations for a range of natural phenomena yet presents other phenomena as anomalies is, all other things being equal, inferior to one that can explain these anomalies. Similarly, an ethical theory that sheds light on some moral intuitions but encourages us to ignore or suppress others is, all other things being equal, less satisfactory than one that does not place this burden on us. This is not to say that the former is thereby shown to be inaccurate. The point is that these omissions show it to be less useful. If what we want from an ethical theory is insight into how to act in ways that convey our moral intuitions—such as that each life is precious, that it is good for people to be happy, that cruelty is bad, that a person's decisions about her life should be respected, and so forth—then a theory that requires us to ignore too many of these intuitions is, on balance, inferior to one that offers plausible proposals for how to manifest them in our actions. A good theory, in short, is one that does not try to *explain away* our intuitions, but that develops proposals for how to convey them in practice.

This formulation of the pragmatist's point may sound abstract, but I take it that Levinas suggests a similar point when he expresses worries about ethical theories in general. I noted in the last chapter that he views ethical theories as double-edged swords. On the one hand, they are valuable tools. On the other hand, we tend to misuse them to legitimate quietism. The danger is that our ethical theories come with "moral blind spots" that allow them to be exploited in this way. For example, Kant's ethics is a useful tool for identifying patterns of behavior that convey the intuition that our ability to make autonomous decisions should be respected. Yet his ethics encourages us to disregard those who lack the cognitive abilities for such deliberations, and thus it invites us to suppress the moral intuition that persons with mental disabilities, the senile elderly, and (arguably) animals deserve our respect. Similarly, one could argue that utilitarianism is a useful tool for identifying actions that voice the moral intuition that it is good to promote happiness, but it has trouble conveying the intuition that persons have rights that should be respected even when doing so does not conduce to the majority's happiness. Neither of these observations entails that these theories are false; they simply tell against their effectiveness. Again, if one thing we desire from an ethical theory is insight into how to enact our moral intuitions, then these theories seem to be limited insofar as they encourage us to ignore certain intuitions. In Levinassian terms, we look to theories of justice for insight into how to convey care for the other in our actions, and a good theory will be one that, among other things, sharpens our capacity to recognize the other, that sharpens our ability to hear her call. By contrast, an ineffective theory will be one that enables us to stifle her claims on us—one that, as I am putting it, *explains away* our moral intuitions rather than *explains them to us*.

Something of this perspective can be seen in Levinas's remarks on Judaism and anti-Semitism. Levinas remarks in several places that Jews have had "the sad privilege to suffer" and associates anti-Semitism with "hatred of the other." Along these lines, consider the following passage from "Antihumanism and Education." Levinas observes that liberal ideals seem discredited to many because they have been misused to defend unjust policies. He adds that Jews are uniquely sensitive to the "fragility of humanism in Western liberalism":

> We, as Jews, were the first to feel it. For us, the crisis of the human ideal, whether of Greek or Roman origin, is announced in anti-Semitism, which is in its essence hatred for a man who is other than oneself—that is to say, hatred for the other man...This martyrdom attests to the fact that the meaning of humanity is not exhausted by the humanists, nor immune to a slippage that is at first imperceptible but can ultimately prove fatal.[43]

Two points interest me in this passage. Levinas claims, first, that Jews' experiences with persecution have sensitized them to how "human ideals" can be misused. Ideals of personhood have too often been used to stigmatize them as nonhuman, and this history has sharpened their sensitivity to the dangers involved in grounding liberal ideals in a substantive conception of personhood rather than in the sheer respect instilled in us by others. He also suggests, second, that Jews hold a special place in modern society as a result of this history. They have had the "sad privilege"[44] to be persecuted, and their "martyrdom" serves as a reminder of how our moral vocabularies can allow "slippages" in our ability to recognize humanity that "ultimately prove fatal."

Levinas seems to think that in every community, there exists some group or groups whose claim on the respect we accord humanity threatens to go unnoticed. Jews have had the "sad privilege" to suffer this role in modern Europe, but one can imagine him granting that other groups occupy this position in other communities. Such ignored groups have a special role to play. They are especially qualified, first, to reveal the mechanisms by which they have been rendered morally invisible. Levinas also seems to think that society redeems itself through them; they serve as a kind of "moral thermometer" of their communities. A society is only as good as its ability to recognize as human, as meriting respect, those whose humanity has been ignored.

Kittay's critique of Rawls gives an example of how these ideas can be deployed against Kantian accounts of justice. Both Rawls and Habermas are interested in identifying fair institutions, but Kittay does a remarkable job of showing how their single-minded focus on fairness blinds them to the humanity of dependents, how it blunts their ability to see the faces of these others. Consider again my example of the parent who petitions the state for a caregiver for her mentally disabled child. The parent's request for a caregiver to improve the child's quality of life is morally invisible to Rawls because the child cannot cultivate the moral powers he associates with citizenship. Yet I take it Levinas would fault Rawls for this failure to recognize the child as having an ethical claim on us. He would see in it an example of that "slippage" in ethical theories which allows "hatred of the other" to take root in the world. This is not to say that Rawls's theory of justice is outright false. Indeed, Levinas would accept it as one of the best tools we have for generating a politics based on the ethical encounter with the other. For the heuristics in his theory, such as the veil of ignorance and the original position, constitute our best models for identifying institutions that fairly distribute resources and opportunities among equally placed individuals, the scenario Levinas associates with "justice" and "the third" in later writings. Levinas's concern with Kantian accounts of justice would rather be that they are not responsive to the full range of moral intuitions we experience in our lives, the full range of intuitions instilled by the other, and

Kittay does an admirable job of exposing that they cannot bear out our moral intuitions about the moral significance of care.

Indeed, Rawls concedes this point. He concludes his line about "normal health care" in *Political Liberalism* with the following admission: "With regard to the problems on which justice as fairness may fail, there are several possibilities. One is that the idea of political justice does not cover everything, nor should we expect it to. Or the problem may indeed be one of political justice but justice as fairness is not correct in this case, however well it may do for other cases."[45] Rawls admits that one might find his account of justice too meager: someone may think that dependents have a right to exceptional care, and it may be beyond an account of justice that takes fairness as its elemental concept to corroborate this claim. He accepts that it is possible to design a more compassionate account of justice; he simply thinks that one rooted in fairness is, for a variety of reasons, superior.[46] Levinas would find this disclaimer unacceptable. He would perhaps accept Rawls's account of justice if it could be shown that justice as fairness is the best we can hope for, but Kittay's criticisms show that the cost of accepting Rawls's theory of justice is high: it requires us to treat a large pool of individuals as ethically invisible in their pleas for additional support, and it places heavy burdens on those who opt to become caregivers.

There is at least one more reason why Levinas would be drawn to Kittay's critique. I complained earlier that Critchley's and Hendley's proposals are unhelpful. To explain Levinas's view of justice, we need to discuss how to record care's ethical significance in general ethical rules, and grafting Kantian tests onto Levinas does not clarify this issue. However, Kittay's aim in critiquing Rawls is precisely to clarify what sorts of social and political institutions do justice to care's ethical significance, and so her criticisms illustrate better the kind of reasoning Levinas would accept. For example, she speculates that it may be necessary to attribute a third "moral power" to one another as citizens:

> If, however, we add the capacity to give care to the other moral powers, and if we include goods related to our interdependence in states of vulnerability in the index of primary goods, we can

> make a case for adding the dependency worker and dependent to the short list of social positions from which to consider the issues of fairness and justice distributions. For example, we can make a case for a paid employee who is in a dependency relation, and so has dependency responsibilities, to receive additional pay, benefits, time off, or services, which would enable her to support the dependency relation in a manner suitable to the situation...This would not be seen as a privilege, but as what is properly due citizens of a just and caring society—enabling each of us to be cared for without extracting an undue burden from those charged with our care.[47]

This proposal is a step in the right direction. Kittay's suggestion is that a community that genuinely values caregiving will include it among the properties it recognizes as vital to citizenship; it will, consequently, subsidize caregiving relations so that they do not fall exclusively on one segment of the population, and it will be responsive to requests among citizens for exceptional forms of care.

Further exposition of Levinas might be needed to show that he would echo Kittay in defending these specific proposals. Nevertheless, my point in invoking her notion of citizens possessing a third "moral power" is simply to illustrate the kind of reasoning he would have us undertake in recording care's significance in a theory of justice. Indeed, he often likens ethical responsibility to the dependency relations that hold between parents and their children. He likens responsibility to "maternity" in *Otherwise than Being*,[48] and in the final section of *Totality and Infinity*, he discusses "fecundity,"[49] the experience of caring for a person who will live on after you and feeling that your life is redeemed through them. One way to understand Levinas is to read him as insisting that it is not only in relation to children, disabled persons, and the frail elderly that we experience dependency. At bottom, my relationship to *each person* is defined by it. Each person should bear the same significance for me as I bore for my parents; my relationship to each should be one of such total responsibility. It is perhaps for this reason that he focuses on cases involving poverty and destitution in discussing face-to-face encounters, for in such situations, other persons are rendered vulnerable to the point where they become akin to dependents.

Kittay does a marvelous job, in turn, of identifying what is required to develop principles of justice that recognize the significance of dependency relations. The fact that such reasoning is absent from Kantian liberal theories suggests, however, that there cannot be a perfect fit between them and Levinas's ethics. This is not to say that Levinas would wholly reject Kantianism, no more than Kittay urges us to reject Rawls's theory outright in proposing a third moral power to supplement the two he discusses. Like Kittay, Levinas would defend Kantian liberalism as an appropriate *starting point* for determining how to enact respect for the other, the most effective tool we have developed for modeling how to fairly adjudicate between conflicting responsibilities. His concern would simply be that it does not go far enough, that it is ineffective in helping us convey certain vitally significant moral intuition in our actions, and thus modifications of the type Kittay proposes should be made to Kantianism so as to make it into a more effective tool.

Moreover, the fact that we can identify these modifications implies that it is possible to generate principles of justice based on Levinas's remarks on the ethical encounter with the other. The criticism with which I began this book, then, seems to be unfounded. Levinas's critics charge that his remarks on ethics are worthless, a "nuisance," because he cannot offer advice on how to put them into practice. Yet he suggests in his writings that Kantianism and Kantian liberalism are good starting points for thinking about how to express the ethical encounter with the other in a theory of justice, and there are grounds for thinking that he would augment this starting point by favoring a more robust account of distributive justice, by insisting that citizens have a right to exceptional forms of health care, and by supporting the idea that caregiving relations should be subsidized by the state. It seems, in short, that Levinas can give a positive account of justice. In this chapter, I have simply tried to model how he might begin to develop such an account by modifying Kantian liberalism. I now turn to explicating in more systematic detail the considerations Levinas would use to develop a positive account of justice and the overall approach to him implied by this book's arguments.

ENDNOTES

1. Critchley, "Derrida and Habermas Get Married," 268.
2. See Critchley, *Cambridge Companion to Levinas,* 27–28.
3. See Hendley, "From Communicative Action to the Face of the Other," *Philosophy Today* 40 (1996): 504–530 and *From Communicative Action to the Face of the Other* (Lanham, MD: Lexington Books, 2000).
4. Levinas likens his ethics to Kant's toward the end of the Substitution section of *Otherwise than Being*, chapter IV, p. 129. He explains in the interview "Being-for-the-Other" that he is sympathetic to Kant's universal law test, although he cautions against the Kantian emphasis on "pure universality." (See Levinas, "Being-for the-Other," 114.) Levinas also speaks favorably of Kant's End-in-Itself in the interview "Reality Has Weight." (Levinas, "Reality Has Weight," in *Is it Righteous to Be?* (Stanford: Stanford University Press, 2001), 158–164. *See esp.* p. 163). Finally, several of his remarks on humanism and human rights, on the impartiality he associates with justice, and his critiques of antihumanist philosophies suggest he is sympathetic to Kant's moral outlook. See Catherine Chalier's *What Ought I to Do?*, trans. Jane Marie Todd (Ithaca: Cornell University Press, 2002) for a thoughtful and thorough overview of the similarities and differences between Levinas and Kant.
5. See Rawls, *Political Liberalism* (New York: Columbia University Press, 1996), 36.
6. Rawls, *Political Liberalism*, xxi.
7. Critchley, "Derrida and Habermas Get Married"; Hendley "From Communicative Action to the Face of the Other," and *From Communicative Action to the Face of the Other*.
8. See Critchley, *Cambridge Companion to Levinas*, 27–28.
9. A further warning may be in order here as well. My remarks on Rawls will focus on his political philosophy as outlined in *Political Liberalism*. Rawls is better known for his *Theory of Justice*, and Critchley refers to this earlier text when he suggests that Levinas's ethics might be integrated into Rawls's political philosophy. I focus on Rawls's later work, however, because he came to feel that the political philosophy he developed in *Theory of Justice* was flawed. *Political Liberalism* represents his more considered point of view.
10. Habermas's and Rawls's exchanges appeared in the *Journal of Philosophy* in 1995. See Habermas, "Reconciliation through the Public Use of Reason: Remarks on John Rawls's Political Liberalism," *Journal of Philosophy* 92

(March 1995): 109–131. Rawls reprinted his "Reply to Habermas" in *Political Liberalism* (New York: Columbia University Press, 1996), 372–434.

11. Rawls, "Reply to Habermas," 380.
12. I do not have in mind any specific texts in the secondary literature but a more general consensus that Levinas's remarks on the specificity and asymmetry of ethical responsibility must put him at odds with Kant. For discussions of Levinas's differences from Kantian ethical theories see Chalier, *What Ought I to Do?*; Hönneth, "The Other of Justice"; Davies, "Sincerity and the End of Theodicy: Three Remarks on Levinas and Kant," in *Cambridge Companion to Levinas*, ed. Simon Critchley and Robert Bernasconi (Cambridge: Cambridge University Press, 2002), 161–187. Several essays on Levinas's relationship to Kant can also be found in *In Proximity: Levinas and the Eighteenth Century*, ed. Melvyn New, Robert Bernasconi, and Richard A. Cohen (Lubbock: Texas Tech University Press, 2001). See, in particular, Anthony Beavers's "Kant and the Problem of Ethical Metaphysics," in *In Proximity*, ed. Melvyn New, Robert Bernasconi, and Richard A. Cohen (Lubbock: Texas Tech University Press, 2001) 285–302 and Peter Atterton's "From Transcendental Freedom to the Other: Levinas and Kant," in *In Proximity*, ed. Melvyn New, Robert Bernasconi, and Richard A. Cohen (Lubbock: Texas Tech University Press, 2001), 327–354.
13. Kant, *Critique of Practical Reason*, trans. Lewis White Beck (New York: Macmillan, 1993), 80.
14. Levinas, *Otherwise than Being*, 56.
15. Levinas, "The Other, Utopia, and Justice," in *Entre Nous*, trans. Michael B. Smith and Barbara Harshav (New York: Columbia University Press, 1998), 223–233. *See esp.* p. 227.
16. Chalier, *What Ought I to Do?*, 39.
17. See, Levinas, "The Paradox of Morality," in *The Provocation of Levinas*, ed. Robert Bernasconi and David Wood (London: Routledge, 1988), 156–167. *See esp.* p. 172. Levinas's difference from Kant on this point can be seen in Kant's argument for charity as an imperfect duty. Kant thinks we have a duty to help others because we cannot will a maxim that advises selfishness to become a law of nature. There inevitably will be moments in my life when I will depend on the kindness of others, and it is inconsistent to excuse myself from a rule I will want others to uphold. Notice the logic of Kant's argument: we should help others because we cannot help but want them to help us. Levinas would object that this ignores precisely what is miraculous about moments when we are moved to treat others beneficently; such moments are remarkable because they show that our concern for others runs deeper than our concern for ourselves. One can imagine Levinas

objecting that the trick is not to show that we should treat others as we would like to be treated. The trick is to make sense of how we can demand significantly greater care for them than we expect for ourselves.

18. See Gilligan, *In a Different Voice* (Cambridge, MA: Harvard University Press, 1982). Another famous example of ethics of care feminism is Nel Noddings's *Caring: A Feminine Approach to Ethics and Moral Education* (Berkeley: University of California Press, 1984).
19. Stocker, "The Schizophrenia of Modern Ethical Theories," in *Virtue Ethics*, ed. Roger Crisp and Michael Slote (Oxford: Oxford University Press, 1997), 66–78. *See esp.* p. 67.
20. For example, see Levinas, "Philosophy, Justice, and Love," 105, 106.
21. See also Annabel Herzog's "Is Liberalism 'All We Need'?: Levinas's Politics of Surplus" for a thoughtful discussion of Levinas's compatibility with liberalism.
22. Schiller, 1940.
23. Seyla Benhabib, *Situating the Self* (London: Routledge, 1992), 187.
24. Hendley, "From Communicative Action to the Face of the Other," 17.
25. See Levinas, "Uniqueness," in *Entre Nous*, trans. Michael B. Smith and Barbara Harshav (New York: Columbia University Press, 1998), 189–196. *See esp.* p. 195, emphasis added.
26. The main place where Critchley challenges this view of deconstruction as apolitical is *The Ethics of Deconstruction*. He first faulted Rorty for his portrayal of Derrida as an "ironist" in "Deconstruction and Pragmatism: Is Derrida a Private Ironist or a Public Liberal?" His essay and a reply by Rorty can be found in *Deconstruction and Pragmatism*, ed. Chantal Mouffe (London: Routledge, 1994), 19–41. Critchley wrote a further reply to Rorty: "Metaphysics in the Dark: A Response to Richard Rorty and Ernesto Laclau," *Political Theory* 26:6 (1998), 803–817.
27. Critchley, "Derrida and Habermas Get Married," 275.
28. Critchley, "Introduction," in *Cambridge Companion to Levinas*, ed. Simon Critchley and Robert Bernasconi (Cambridge: Cambridge University Press), 1–32. *See esp.* p. 28. Critchley borrows this distinction between legislators and perfectionists from Hilary Putnam's essay "Levinas and Judaism," 36–43, and Putnam borrows the distinction, in turn, from Stanley Cavell's *Conditions Handsome and Unhandsome*.
29. I should note in fairness to Critchley that he has not continued to defend this pairing in his more recent publications. He proposes wedding Levinas with Kantian liberals in his Introduction to *The Cambridge Companion to Levinas* and in "Derrida and Habermas Get Married." However, he

has defended a more anarchistic interpretation of Levinas in his recent publications, such as *Infinitely Demanding* and "Five Problems in Levinas's View of Politics." I address this anarchistic interpretation in closing pages of chapter 3, footnote 135.

30. My claim against Hendley is fairly minor. I am largely in agreement with his observation that Levinas rejects partialist approaches and with his argument that Levinas accepts a broadly Kantian ethical framework. Indeed, I offered my own reasons for this interpretation in the last chapter. My disagreements with Hendley, as I explain later, are threefold. First, documenting that Levinas favors impartialism does not itself establish that he accepts Kantianism, for consequentialist theories are also opposed to partialism. Second, it seems to me that Hendley's proposal amounts to simply grafting Kant onto Levinas rather than showing that Levinas would reason his way to Kantian norms. His proposal does not clarify how to generate an account of justice based on Levinas's remarks on the ethical encounter with the other. Finally, the fit between Levinas and Kant is not perfect. When we try to determine what moral norms Levinas would endorse, there are reasons to think he would attach greater significance to caregiving than would Kantians such as Rawls and Habermas. I accept Hendley's basic point, then, that Levinas starts with a broadly Kantian ethical framework, but I also think Levinas would modify this framework, favoring a more robust account of distributive justice than most Kantians accept.
31. Critchley, *Ethics of Deconstruction*, 268.
32. See Levinas, "Reality Has Weight," in *Is it Righteous to Be?*, ed. Jill Robbins (Stanford: Stanford University Press, 2001), 163.
33. See Singer, "Famine, Affluence, and Morality," *Philosophy and Public Affairs* 1:3 (1972): 229–243, and *Practical Ethics* (Cambridge: Cambridge University Press, 1979).
34. Critchley, "Derrida and Habermas get Married," 269.
35. Levinas, *Otherwise than Being*, 159.
36. See Rawls, "Kantian Constructivism in Moral Theory," *Journal of Philosophy* 77:9 (September 1980): 515–572, *see esp.* p. 526; *Political Liberalism*, 181. My list of Rawls's primary social goods is a composite of the versions given in these two sources. Rawls's list of primary goods remained the same throughout his career; each of these sources simply uses a slightly different wording.
37. Kittay, *Love's Labor* (London: Routledge, 1999), 76.
38. Ibid., 113.
39. Rawls, *Political Liberalism*, 21.

40. Ibid., 184.
41. Ibid.
42. By "day-to-day health care," I have in mind occupations such as social workers, nannies, and live-in caregivers for elderly persons—not doctors or nurses. The salaries for these positions tend to be low, and they tend to come with relatively few benefits.
43. Levinas, "Antihumanism and Education," in *Difficult Freedom*, trans. Seán Hand (Baltimore: Johns Hopkins University Press, 1997), 277–288. *See esp.* p. 281.
44. Levinas, "Antihumanism and Education," 281.
45. Rawls, *Political Liberalism*, 21.
46. Rawls's main argument for this claim in *Political Liberalism* is that "justice as fairness" is the easiest account of justice to defend. As I noted earlier, Rawls worries that forms of liberalism that rely on substantive truth claims, so-called "comprehensive liberalisms," are flawed because their acceptance hinges on whether one accepts the truths on which they are premised, whereas Rawls hopes to establish a "freestanding" theory of justice that can be defended to anyone because it does not rely on *any* truth claims for its justification. Rawls would probably worry, then, that Levinas's account of justice, one that seeks to develop norms based on care, reduces to a comprehensive liberalism that cannot be justified to all skeptics.
47. Kittay, *Love's Labor*, 112.
48. Levinas, *Otherwise than Being*, 67, 71, 75, 76.
49. See Levinas, *Totality and Infinity*, 276–270.

Chapter 5

Levinas as Pragmatist

Those who have worked on methodology all their lives have written many books that replace the more interesting books that they could have written. So much the worse for the philosophy that would walk in sunlight without shadows.

—Levinas "Questions and Answers"

1. Introduction

Let me take stock of where my argument has left us. My overarching aim has been to defend Levinas against the criticism that he cannot offer his readers an account of justice. We saw in chapter 1 that this worry turns on a puzzle concerning the relationship between the theoretical and practical sides of his philosophy. The worry makes sense if we read him as starting with an epistemic observation about the inexpressibility of the relation to the other and moving to the claim that we must nonetheless express this relation in a system of moral norms. We saw in chapter 2, however, that he should be read the other way around: he begins with the ethical insight that we cannot help but acknowledge one another as overridingly valuable and concludes from it that it can be misleading to portray this acknowledgment in cognitive-epistemic terms. I suggested that nothing in this alternative position prohibits him from developing an account of justice, and, indeed, that hints of one can be gleaned from his writings. My task in the past two chapters has been to make good on these claims by, first, drawing attention to areas where he indicates what account of justice he favors, and, second, in chapter 4, by modeling how Levinasian modifications can be introduced into Kantian accounts of justice.

I now want to give a more systematic overview of the procedures Levinas would use to develop a positive account of justice as well as the overall approach to him I have sought to develop through this book's arguments. I would characterize this approach as "pragmatic," a term I use to call to mind American philosophers such as William James, John Dewey, and, more recently, Richard Rorty and Hilary Putnam. I will have more to say later in this chapter about why I find it useful to liken Levinas to the pragmatists. It will be helpful, however, to begin by outlining the alternative schools of thought.

2. Empirical, Transcendental, and Deconstructionist Readings of Levinas

Robert Bernasconi's "Rereading *Totality and Infinity*" offers perhaps the best overview of the main interpretations of Levinas in the secondary literature on him. Bernasconi begins his 1989 essay by considering what "habits of reading" have developed since the publication of *Totality and Infinity* in 1961. He suggests that two schools of thought have emerged on the role of the face-to-face relation with the other in Levinas, what he calls the "empirical" and "transcendental" interpretations. He adds that these approaches "provide the terms for almost every introduction to Levinas's thought." He goes on to raise worries about both, worries that proved prescient. Levinas scholars who were influenced by Bernasconi, such as Simon Critchley, have amplified on them in arguing for what can be understood as a third, deconstructionist reading. Bernasconi's essay is valuable, then, because it neatly articulates what seem to be the three major schools of thought about Levinas.

Let me begin by saying something about what Bernasconi calls the "empirical" and the "transcendental" readings. Each of these readings focuses on Levinas's remarks on the face-to-face encounter and the extent to which the relation to the other is expressible in our day-to-day lives. "Some interpreters understand it," he writes, "as a concrete experience that we can recognize in our lives. Other commentators have understood the face-to-face relation to be the condition for the possibility of ethics and indeed of all economic existence and knowledge."[1] Bernasconi calls the first approach the "empirical reading" of Levinas and the second the "transcendental reading."

My remarks in this book may lead you to think I favor the empirical reading. My examples of face-to-face encounters tend to be discrete episodes in one's life when one witnesses a person in a position of vulnerability and feels pity for her. My stock examples are watching news reports where one sees people victimized, passing by someone begging for help, and the scenes from Grossman's *Life and Fate*. Based on these examples, I have said several times that a face-to-face encounter is a moment when we become keenly aware of another person's humanity, with the caveat that what these moments reveal is that another person is, first and foremost, a source of responsibility. So, I seem to support something like the empirical reading. This reading takes Levinas to be suggesting that the relation to the other can be expressed in our everyday lives, and, indeed, that face-to-face encounters are moments when this relation becomes conspicuous.

There are problems with this approach. One problem should be obvious given how often I have harped on the ineffability hypothesis. Levinas repeatedly says that the other "overflows" efforts to comprehend her; she is infinite, transcendent, uncontainable, absolutely other, and ungraspable. Given such remarks, it seems like it should be impossible to have any direct experience of the other in our lives. Moreover, Levinas often depicts the relation to the other as a kind of primordial precursor to the relations and concepts that define our everyday mode of being in the world. The relation to the other is *prior* to consciousness, *prior* to being, and *prior* to language. One of Levinas's mantras is that the relation to the other is prior to ontology, prior, that is, to the gamut of objects, concepts, life experiences, and sensations that define our everyday mode of engagement in the world. However, if the relation to the other is prior to ontology, if it is preconceptual, prelinguistic, and preexperiential, then it should be impossible to have any direct, concrete experience of the other's presence.

Here it is worth noting that Levinas's remarks on the infinitude and primordiality of the relation to the other do not seem to be mere rhetorical flourishes. He once admitted that he frequently uses hyperbole to jar readers out of conventional ways of thinking.[2] However, he goes to great lengths to explicate the subtle ways in which the relation to the other cannot be rendered in terms of the concepts we use to reflect on everyday life experiences. For example, he notes in several places how Husserl's notion of intentionality, the idea that all conscious perception has some object, falsifies the relation to the other. The relation to the other cannot be represented, he seems to argue, in the "meaning structures" that define our standard spatiotemporal mode of experiencing the world. The other is not, as he puts it in a 1965 essay, a "phenomenon" but an "enigma."[3] Elsewhere, he suggests that one's relation to the other is not an object of experience but a "curvature of intersubjective space."[4]

These kinds of worries have led some to accept the transcendental reading of Levinas. On this view, the relation to the other is not a concrete event we recognize *in* our lives but a *condition* of them. The reference here is, of course, to Kant's transcendental arguments. There is a massive secondary literature, both in analytic and in continental philosophy, on transcendental arguments. Charles Taylor offers a helpful summary, however, of their key characteristics:

> The arguments I want to call "transcendental" start from some feature of our experience which they claim to be indubitable and beyond cavil. They then move to a stronger conclusion, one concerning the nature of the subject or the subject's position in the world. They make this move by a regressive argument, to the

> effect that the stronger conclusion must be so if the indubitable fact about experience is to be possible.[5]

Typically, this type of argument has been used against skeptics. The skeptic questions the reality of, say, causality. One seeks to show in a transcendental argument that causality is implied by something the skeptic cannot question, such as the fact that we are conscious at all. Roughly speaking, a transcendental argument tries to clarify the conditions for the possibility of a domain of phenomena that is "beyond cavil," the conditions of, for example, consciousness, perception, language, agency, or meaningful social interaction.

The transcendental approach interprets Levinas, then, as attempting to show that the relation to the other is something like a condition for rational deliberation, consciousness, language, ethics, and so forth. At least two factors tell in favor of this approach. The first is that there is textual evidence for it. Let me confine myself to Levinas's main argument in *Totality and Infinity*. Levinas subtitles *Totality and Infinity* an "essay on exteriority." "Exteriority" is, roughly, his term for something like mind-independent reality, for whatever is "exterior" to the self's experience of the world and itself, whatever is "exterior" to what can be known, comprehended, or perceived. Levinas uses "interiority" to refer to the gamut of concepts and relations that define the self's experience of the world. One of his key observations in the preface and section I of *Totality and Infinity* is that the elemental categories that define interiority assume a relation to exteriority that they cannot themselves supply. Interiority has a "transcendent intention."[6] The thesis he proposes to defend is that it is through the responsibility we feel for others that we brush up against something wholly exterior to interiority. Thus he explains in his preface that he plans to "proceed from the experience of totality back to a situation where totality breaks up, *a situation that conditions the totality itself.* Such a situation is the gleam of exteriority or of transcendence in the face of the Other."[7]

There is textual evidence, then, that Levinas saw himself as engaged in something like a transcendental analysis of the conditions of the possibility of experience. Indeed, he expressly uses the language of transcendental arguments in *Totality and Infinity*. "For the way we are describing to work back and remain this side of objective certitude," he writes, "resembles what has come to be called the transcendental method."[8]

The second factor that tells in favor of this approach is that it avoids the flaws of the empirical approach. The transcendental approach can accept Levinas's claims about the infinitude and primordial character of the relation to the other, for it does not depict this relation as a recognizable event *in* our lives but an *underlying* structure that shapes them in their entirety. Proponents of this approach have been careful, too, to insist that the relation to the other cannot be directly *deduced* by reflecting on the self's experience of the world, hence it cannot be assimilated back into the domain of interiority.

Thus Theodore De Boer, the main proponent of this approach, argues in "An Ethical Transcendental Philosophy" that Levinas's aim is to show how "the philosophy of the Other is able to provide a foundation" for "phenomenological ontology."[9] Yet he cautions that this "transcendental condition is not a necessary ontological structure which can be recovered from the empirical phenomena; it is an unrecognizable event or ontic incident which intersects the ontological order."[10] De Boer's suggestion is that interiority, or phenomenological ontology, the domain of the self's experience of the world, presumes a relation to exteriority. The relation to the other is thus something like a condition for perception, comprehension, and deliberation. However, the mere fact that thought and perception are directed to something beyond themselves, to the other, does not mean that this something is, so to speak, really out there. Yet we do undergo unique moments in our lives that resist comprehension, "unrecognizable events," involving our relationships to others that, in a way, bear out the presence of this something transcendent. The relation to the other is thus both a condition for consciousness yet not directly derivable from it.

Levinas is similarly cautious in *Totality and Infinity*. He proposes to show that the relation to the other is a kind of condition for interiority, but warns that the "revelation of infinity" that occurs through the other's face does not "lead to the acceptance of any dogmatic content."[11] Elsewhere he explains that his goal is to show how "objectifying thought" is "sustained" by a "forgotten experience from which it lives" but warns that this "deduction" is "necessary yet non-analytical."[12] Even in describing his main argument as transcendental, he is careful to stress that his method merely "*resembles* what has come to be called the transcendental method."[13]

So, what is wrong with the transcendental interpretation? One problem is that there seems to be a fair amount of textual evidence for the empirical reading as well. Levinas does talk about face-to-face encounters as if they were concrete events. He is unambiguous in citing the two scenes from *Life and Fate* as examples of the "welcoming of the face," and these scenes seem to involve recognizable events in the lives of the novel's characters. The empirical reading may jibe more, too, with Levinas's ethical-political insights. Levinas objects to Nazism and Stalinism on the grounds that these political ideologies inhibit their adherents from recognizing the preciousness of each individual's life. This seems to imply that at various moments in our lives, we can do a better or worse job of recognizing the other and our responsibility for her. The empirical view seems to be correct, therefore, in its claim that the relation to the other is a concrete experience that we can recognize in our lives, or, alternatively, that can be stifled or absent from them under certain political regimes. By contrast, the transcendental approach seems to lead to the conclusion that respect for the other is present even in Nazism, for it maintains that the relation to the other is an underlying structure that shapes all facets of our lives.

Most importantly, the transcendental reading is ambiguous. On the one hand, it contends that the relation to the other is a condition for interiority: my world could not be the way it is if I did not stand in relation to something wholly exterior and irreducible to it. On the other hand, it insists that I cannot, properly speaking, *deduce* this "something wholly exterior" as a condition of my world; the relation to the other is not, de Boer insists, "a necessary ontological structure that can be reconstructed from the empirical phenomena." So, there is an ambiguity about how interiority and exteriority fit together. Thus Bernasconi notes how de Boer waffles in his characterizations of the relation to the other, referring to it as an experience, yet insisting that it cannot be an experience.[14] De Boer initially argues that for Levinas, "the transcendental condition is an ethical *experience* enacted in dialogue." Yet later he remarks that "the condition for the possibility of experience is not experience itself."

It is this ambiguity that leads Bernasconi to favor a more deconstructionist reading of Levinas in "Rereading *Totality and Infinity*." Bernasconi's critique of de Boer and his analysis of the section II of *Totality and Infinity* are subtle and rich. However, I want to pass over some of their details. What interests me is the general approach to Levinas's philosophy that he defends in this essay, an approach that others have gone on to elaborate on and defend. Let me offer a barebones sketch, then, of this deconstructionist approach.

Bernasconi notes that evidence for both the empirical and the transcendental readings can be found in Levinas. "Levinas himself," he writes, "seems to be unable to decide between these rival interpretations."[15] Some interpreters stress passages where Levinas depicts the face-to-face encounter as a concrete episode; others stress passages that support the transcendental interpretation. Bernasconi's provocative suggestion is that it is a mistake to favor either. Levinas is best read as staging a "displacement" of both:

> It seems to me that Levinas is using the language of transcendental philosophy and the language of empiricism...in an effort to find a way between these twin options given to us by the philosophical... language we have inherited. Only by employing both languages and drawing them into contradiction as he does...can he hope to introduce us to a way of thinking which rests on neither.[16]

To read Levinas this way would be to read him as a kind of deconstructionist, as sharing in the broadly deconstructionist idea that philosophic thought has been structured around various binary oppositions, that we cannot simply transcend these oppositions, but that we can begin to move beyond them by exposing how they are contradictory and self-refuting. Thus Bernasconi notes how Derrida, in "Violence and Metaphysics," exposed the ways in which these two sides of Levinas's philosophy seem to continually undermine each other. He asks, however, whether Derrida might not have underestimated Levinas, whether "Levinas was unsuspecting…as Derrida, in places in [his] essay, seems to suggest."[17]

As I noted earlier, Bernasconi's argument for this approach is complex, resting on a subtle exegesis of section II of *Totality and Infinity*, and I want to pass over its details so I can clarify his overall reading itself. Bernasconi presents his deconstructionist interpretation as a critique of de Boer. He can be read, however, as offering a subtler analysis of the same ambiguities we saw at play in de Boer. One of the problems with the transcendental approach, we saw, is that it is ambivalent: it presents the relation to the other as both an experience and a nonexperience, as a condition for experience and a noncondition. The deconstructionist approach embraces these contradictions: Levinas's point is precisely to show how our thinking is structured by a relation that we cannot comprehend and that, consequently, generates these contradictions. The difference between this approach and the transcendental approach is that the deconstructionist does not view Levinas as *deducing* conditions but as exposing a necessary aporia in the concepts and languages we use to make sense of our lives. Recall de Boer's claim that the relation to the other, as a transcendental condition, is "not a necessary ontological structure" but an "unrecognizable event." Similarly, the deconstructionist takes Levinas to be exposing how our thought is structured by a relation it can never truly grasp. His aim is not to offer some philosophic theory of his own about the other, or even to directly prove that the other is a condition of thought or experience, but to deconstruct the two major philosophic discourses that have come to shape our thinking, the discourses of empirical and transcendental thinking, by drawing them into contradiction with each other and, in doing so, indirectly showing that they are influenced by a relation that neither can fully capture.

I want to assume for the sake of argument that this deconstructionist approach is superior to the empiricist and transcendental interpretations.

There are subtle versions of both interpretations.[18] Nonetheless, I find it useful to assume that the deconstructionist's criticisms are sound. First, this approach matches the trajectory in the scholarship on Levinas. A broadly deconstructionist reading of Levinas seems to have become well accepted in recent scholarship, and it seems to have replaced the earlier schools of thought. Second, this approach is also the foil to my own interpretation. More than the other two, it embraces the cluster of assumptions I associate with the ineffability hypothesis. Indeed, the gist of its criticisms of the other interpretations is precisely that they disregard the inexpressibility of the face-to-face relation: the empiricist mistakenly assumes it can be expressed in a discrete experience, and the transcendental interpretation assumes it can be derived from experience. The deconstructionist approach, then, represents the kind of approach I have sought to call into question by disputing the ineffability hypothesis.[19]

3. A Pragmatic Reading of Levinas

What alternative approach to Levinas is suggested by the defense elaborated in this book? I would characterize my approach as pragmatic.

In calling it "pragmatic," I mean to call to mind the everyday definition of this term as "active rather than contemplative" and as "pertaining to the study of events… with emphasis on their practical consequences." My aim is also to associate Levinas with American pragmatism, with figures such as James and Dewey and, more recently, Richard Rorty and Hilary Putnam.[20] In comparing Levinas to these figures, my intent is not to suggest that his thinking is exactly the same as James's or Dewey's, but merely to suggest that it shares broad similarities with insights propounded by the pragmatists. I want to be careful, though, in suggesting these similarities. A frustrating trait among admirers of pragmatism is their tendency to paper over the differences between dissimilar philosophers, so that one stumbles across lines in which James, Dewey, Nietzsche, Wittgenstein, Heidegger, Foucault, Kuhn, and Derrida are all uttered in the same breath, as if their works were interchangeable. I want to avoid extending the term "pragmatism" to the point where it becomes empty, so let me stress that there are important differences between Levinas and recent pragmatists, and I will elaborate on some of them once I have explained why I find it useful to compare him to pragmatists in the first place.

There are at least two useful points of comparison between Levinas and pragmatists. First, they uphold similar epistemic stances in linking broadly epistemological questions about truth and meaning with practical consequences. Both contend that epistemological claims about whether consciousness, language, or conceptual schemes correspond to reality are not so much fallacious, a skeptical thesis, as they think that debating about such issues is counterproductive and that a better test is whether they help us solve real-world problems. Along these lines, Richard Rorty explains in "Solidarity or Objectivity?" that his brand of pragmatism is not a positive but a negative position: it does not claim that veridical knowledge of mind-independent reality is impossible, but that debating about it is counterproductive. We would be better served if we put aside the question of whether our descriptions of the world correspond to its true nature and focused instead on assessing whether they help us resolve crises, accepting them as true, for all intents and purposes, when they meet this condition.

Levinas writes from a similar perspective. His claim is not that it is impossible to comprehend the other but that it is misguided to treat her as an object of comprehension. It is inappropriate because my relationship to the other is defined by responsibilities. The other is, first and foremost, as I noted in chapter 2, a kind of living manifestation of the biblical prescription against murder: *Thou shalt not kill!* That is, the other confronts me as a source of unconditional demands: *Do not hurt me. Allow me to live. Allow me, too, to draw nourishment from this world.* This source of demand, in turn, requires me to grapple with a series of challenging questions: *Am I in any way harming or neglecting my responsibilities to this other? What should I do to help? Have I done enough? Can I do more?* Levinas's point in claiming that the other cannot be comprehended is not mystical or obscurantist; his point is that this relation is defined by obligations to help, and whether I have grasped this fact is not reflected in how I *discuss* it, in claims to *know* it, but in the *deeds* I perform to help.[21]

The more general point he makes is that the substance of our lives is defined by such agent-focused questions. He finds it useful to talk about singular encounters with the other, passing by a stranger in need of help, but really we are immersed in a sea of obligations. I am duty bound to care for my family, but I am also duty bound to care for strangers. I am responsible for the suffering of workers in my home country when their jobs are shipped overseas, but I am also responsible for the suffering of workers in other countries, both when they are paid subpoverty wages for these jobs as well as the additional poverty they undergo when, as a result of protectionist economic policies in my home country, they are deprived of even this meager income. The entirety of my life is awash in such difficult questions of justice. This last insight leads Levinas, I argue, to follow the pragmatists in linking broadly epistemological issues about truth and meaning to practical consequences. Levinas accepts that we can make legitimate epistemological claims about the world, ourselves, and others. However, we are always faced with more pressing dilemmas, and so, the pivotal question with respect to our theoretical descriptions of reality is not primarily whether they are accurate but whether they help us resolve the crises that beset us.

Levinas and the pragmatists come to similar conclusions, second, about politics and normative ethics based on this prioritization of what Putnam and Rorty call the "agent point of view" over the "spectatorial point of view." A certain reformist sensibility is common to both. This is not to say that either is wholly opposed to political revolution or to dramatic, large-scale social changes. For both, however, a consequence of rejecting the primacy of the spectatorial point of view is that a certain justification for political change can no longer be accepted.

Rorty discusses this point in "Solidarity or Objectivity?," where he distinguishes "realist" from "pragmatist" approaches to politics. According to Rorty, most political philosophies have been realist in their efforts to "ground solidarity in objectivity."[22] Most political philosophies have championed programs of social reform on the grounds that these programs correspond to some ahistoric truth, typically an ahistoric truth about human nature. Democratic institutions should be accepted as the ideal form of government, for example, the realist argues, because they correspond to the fact that all humans possess free will. In short, the realist urges *solidarity* for a political scheme by attempting to show that it is *objectively* true. By contrast, the pragmatist tries to "reduce objectivity to solidarity." They argue that debate over ahistoric truths is a pointless enterprise, and they contend that a better goal is to aim for "as much intersubjective agreement as possible." Thus the pragmatist might argue for liberal, democratic institutions, not on the grounds that they reflect a deep truth about human nature, but because of the respect they accord to individual freedoms and their low tolerance for cruelty. For the pragmatist, a political scheme that inspires *solidarity*, that reflects the best solution we mutually develop in response to the crises that confront us, should be accepted, for all intents and purposes, as *objective*.

This rejection of realism leads the pragmatist to embrace a more reformist sensibility. Traditionally, realists have been able to critique their governments for being out of synch with one ahistoric truth or another, and they have argued on behalf of revolutionary change by contending that it is needed to bring their governments in line with them. Thus the orthodox Marxist could argue for revolution on the grounds that it was needed to achieve the classless society that was the outgrowth of capitalism. However, the pragmatist contends that it is counterproductive to gauge social and political institutions against such ahistoric truths. What matters is not whether one's institutions *correspond* with some platonic form of the ideal state, final end goal of history, or deep truth about human nature. What matters is whether they work: Are they accepted by people as constituting the best solutions available to them for the problems they face? For the pragmatist, in turn, there is a sense in which all politics is local and reformist. That is, all political change, even if massive in scope, should express what citizens of a community envision as the best way to modify their existing social and political institutions in the face of the very particular challenges they hope to resolve.

Levinas's denial of spectatorial standpoints in philosophy leads him to accept a similarly reformist position. Indeed, this claim constituted a key step in my argument in chapter 3. There I argued that Levinas's reluctance to elaborate a normative ethics should not be construed as evidence that he finds it impossible to do so but of his pragmatism. What Levinas would call "questions of justice" are not questions about what political institutions are objectively true, but questions about how to negotiate the conflicting demands we experience in our responsibility for each other. I also suggested that it follows that he views accounts of justice and normative ethical theories as tools we use to attune ourselves to one another's needs, and he is reluctant to advance any single ethics because flexibility is needed to discern which ethics best enables us, in a given situation, to render ourselves maximally responsive to one another.[23]

Let me offer further evidence, however, beyond what I covered in chapter 3. Levinas's sympathy for the pragmatist's reformism can be seen in his criticisms of totalizing ideologies, particularly his critique and corresponding praise for "eschatology" in the opening pages of *Totality and Infinity*. Levinas opens *Totality and Infinity* by asking a question of "highest importance": "whether we are not duped by morality."[24] He suggests that most political ideologies, and most philosophic perspectives, assume that genuine "morality" and "peace" are impossible. We can at best achieve something like morality by "founding morality on politics."[25] Levinas is a little unclear in these opening pages about what he means by "morality." His point seems to be this. Most political ideologies encourage us to view other individuals as mere "bearers of forces that command them unbeknown to themselves." I am led to view other persons as instantiations of broader historical-political processes, such as socioeconomic processes, my tribe's cultural values, God's will, the manifest destiny of my nation, a generic rights-bearing agent, and so forth. This view leads to a picture of communities as wholes composed of competing, antagonistic forces. To found morality on politics would be to try to establish a kind of peace within this tension by achieving equilibrium among these forces.

By contrast, Levinas associates genuine morality with eschatology. "Eschatology" is typically used to refer to the branch of theology concerned with the final events in the history of the world. Levinas uses this term somewhat differently:

> The eschatological, as the "beyond" of history, draws beings out of the jurisdiction of history and the future; it arouses them in and calls them forth to their full responsibility. Submitting history as a whole to judgment, exterior to the very wars that mark its end, it restores to each instant its full signification in that very instant: all the causes are ready to be heard.[26]

Levinas uses "eschatological" to refer, then, not to the *endpoint* of history, but being *uprooted from history as such*, being placed outside it and called to cast judgment upon it. What does it mean to cast judgment on history? Levinas has in mind something like the standpoint Ivan Karamazov asks Alyosha to contemplate in the famous "Rebellion" section of *Brothers Karamazov*. Ivan asks Alyosha to imagine standing at the end of time and learning that a single innocent's suffering played a pivotal role in a divine plan that has culminated in universal happiness. Would Alyosha accept this suffering as justified? Notice that Alyosha is here faced with a kind of criterionless choice. The question is not whether the innocent's suffering is *necessary* to fulfill some divine plan; Ivan asks Alyosha to imagine that this question is settled. Instead, he invites Alyosha to submit God's plan and the whole of history itself to judgment. Standing beyond history, seeing its fulfillment, could you content yourself with this state of affairs, or would you rage against it as unjust?

Levinas was a careful reader of Dostoevsky, and he seems to be making a similar point in his remarks on eschatology about how witnessing suffering uproots us from history. His insight is that we are continually being drawn out of history in the way Ivan envisages. I may wholeheartedly believe that free markets and deregulated labor are the best way to improve quality of life, or, conversely, that happiness can only be achieved by abolishing private property. Or I may firmly believe that secular values must be spread to the rest of the world, or that the only fulfilling way to live is through service to God. However, *whatever* I believe, I am forced to radically interrogate *all* such beliefs every time I witness someone in need. I am drawn out of history—forced to weigh all such grand goals against the harm done to the particular individual before me. Levinas here sees a break between the standpoint of politics—and by extension most of philosophy—and the eschatological perspective of true morality. For, from the standpoint of morality and eschatology, I no longer see the individual as a "bearer" of larger, anonymous forces; their need leads me to view them as a unique individual whose plight calls on me to subject history itself to judgment, to, as Levinas puts it, restore to "each instant its full signification."

More could be said about the opening argument of *Totality and Infinity*. I want to focus, however, on the general approach to ethics and politics that flows out of it. Levinas's concern is that people too often ignore the particular concerns of individuals by viewing them as mere *dramatis personae* in larger social, historical, or political processes. In short, they totalize: they disregard the other's particularity and reduce her to an instantiation of a general, abstract "totality." By contrast, Levinas thinks that the needs of particular individuals must always be primary, and the only justification for what he later terms "justice" is that it helps us address these needs. Like the pragmatist, then, there is the idea in Levinas that the focus of ethics and politics should be on developing imaginative ways to solve the particular crises that beset particular individuals, and there is a similar rejection of approaches to ethics and politics which are realist in their efforts to justify social and political changes on the grounds that they are needed to bring our institutions into line with ahistoric truths. Just as the pragmatist would have us eschew debates over whether our ethical theories are *true*, and focus on whether they are *effective* in resloving the crises that beset us, so too Levinas rejects the idea that the task of ethics is to bring our practices in line with some final end goal of history, be it some platonic form of the ideal state, classless society, or deep truth about human nature. Instead, my responsibility for those concrete "others" with whom I share the world moves me "beyond history"; it requires me to gauge my practices based on whether they do, in fact, lead me to aid these others: Do they supply me with effective solutions to the conflicts that beset them? Does it aid me in resolving conflicts between them? Levinas defends a similarly reformist position as the pragmatist, then, in that he too insists that the fundamental criteria we should use in assessing our practices is whether they constitute our most imaginative solutions to the particular crises that beset particular individuals.

Here, though, I need to be careful. I noted that there is a tendency in discussing pragmatism to paper over its differences from other schools of thought. I am arguing for a pragmatic approach to Levinas, but I do not wish to claim that he *is* a pragmatist. So, what are some differences between his perspective and that of pragmatists?

The most significant difference is that they offer different justifications for rejecting the primacy of the spectatorial standpoint. Both link epistemic questions about truth and meaning with practical consequences. However, contemporary pragmatists draw this connection from within epistemology, whereas Levinas establishes it by reflecting on ethical responsibility itself. The pragmatist reaches her conclusions through epistemological arguments about the nature of language, justification, or scientific theory, whereas Levinas's arguments are metaethical. For example, a pragmatist might argue, apropos of Kuhn's *Structure of Scientific Revolutions*, that theoretical descriptions of reality are regarded as sound in the scientific community not because they have been conclusively proven to correspond with reality but because they possess a variety of virtues, such as being parsimonious, having broad explanatory range, and being able to solve a range of problems more efficiently than do the theories they replace. Pragmatists tend to argue, that is, that philosophers have misunderstood the mechanisms by which we construct knowledge: it is not achieved by matching our descriptions of the world up with mind-independent facts but by formulating observations that cohere within broader systems of beliefs that, in turn, help us solve real-world problems efficiently.

By contrast, Levinas reaches his conclusions by reflecting on the nature of ethical responsibility, and his rejection of spectatorial standpoints is metaethical rather than epistemic. His central concern is with the finality of moral obligation—the way the moral *ought* overrides other considerations. He begins with the insight that we cannot help but acknowledge one another as morally valuable, which is to say *overridingly* valuable. It follows that our moral obligations to one another cannot help but pervade all aspects of our lives. Faced with the choice between personal ambitions and my duty to the other, I am obligated to place the other before me. The effect of this point is to render the spectatorial point of view *morally illicit*. It may be personally enriching for us to engage in the disinterested pursuit of knowledge, but the only *morally satisfying* justification for it, in the face of the suffering of others, is that it ultimately helps us better attend to their needs.[27] Simply put, whereas the pragmatist argues that we only accept knowledge claims as justified when they help us solve problems of significance, Levinas's point is that it is irresponsible to busy ourselves with knowledge claims that lack any significance in a world where so many suffer so dramatically.

This alternative critique of the spectatorial standpoint leads Levinas to be even more pragmatic, arguably, than recent pragmatists. Richard Rorty argues time and time again that philosophers should abandon their efforts to uncover an unshakeable foundation for knowledge and focus on practical issues involving politics and day-to-day life. Philosophers have critiqued him by arguing that his antifoundationalism commits him to an irrational form of relativism and that his criticisms of foundationalism are unsound. At times, he has replied by embracing this critique, claiming it would be contradictory to attempt to refute it since it would require him to employ "Platonist" assumptions he rejects. However, his arguments for antifoundationalism court these replies. His argument in *Philosophy and the Mirror of Nature* borrows from Wilfrid Sellars's critique of the "myth of the given," and *Contingency, Irony, and Solidarity* makes heavy use of Wittgenstein's and Davidson's ideas about language. So, it is understandable that philosophers would return to epistemic questions about language and certainty rather than follow Rorty's recommendation that they set them aside, for Rorty himself invokes them in arguing for this proposal. The effect is that there is something mildly contradictory about his writings: for all their talk about abandoning pointless debates over foundationalism, they devote a lot of time to discussing the very problems which, they contend, it is pointless to discuss.

Levinas does not face this problem. For his ground for rejecting the spectatorial standpoint does not rely on antifoundationalist epistemic critique. His thesis is *not* that it is impossible to know the other. As controversial as it may sound, it is my belief that he accepts that we can make all sorts of licit knowledge claims about the other. I can know that this particular other's eyes are, say, brown, that she has two hands, that she was a student in my ethics class two years ago. Levinas's point is not that I am *wrong* when I make such judgments, that I have not *really comprehended* the other, but that these knowings do not get at the root of my relationship to the other. His corresponding worry about philosophers is not that they are wrong to want a foundation for knowledge, but that this impulse is dangerous insofar as it enables a uniquely pernicious form of quietism, insofar as it feeds into our efforts to shirk our responsibilities by misrepresenting our relationship to others.

Indeed, there is something of the pot calling the kettle black in Rorty's tirade against Levinas. Rorty complains that Levinas's writings, for all their bluster about ethics, have no practical significance and distract attention away from more useful discussions about how to solve real-world political dilemmas. Yet this charge can be leveled against Rorty: for all his talk about putting aside foundationalist debates, he harps a great deal about foundationalist issues, such as why it is mistaken to think that our ideas correspond to some fundamental reality. This last criticism may be somewhat unfair. Rorty published numerous pieces, particularly in later works, where he reflects on contemporary political dilemmas. He could argue as well that he could not discuss these issues constructively had he not elaborated his antifoundationalism—that had he not offered these critiques any discussion would be hijacked by our tendency to adopt "realist" positions in our ethical and political thinking. However, Levinas can offer the same mitigating consideration. He too can argue that he needed to focus on elaborating his critique of philosophy as totalizing in order to clear the way for a constructive discussion of matters of ethical significance.

4. Defending the Pragmatic Approach

Earlier, I used Robert Bernasconi's "Rereading *Totality and Infinity*" to outline the three main schools of thought in the scholarship on Levinas, and I offered that the deconstructionist is the most sophisticated of the three. This approach is also the foil to my pragmatic interpretation, which denies that Levinas is committed to controversial epistemic claims about the inexpressibility of the other. Instead, it reads him as critiquing what I, following Rorty and Putnam, call the spectatorial standpoint, the temptation in philosophy and in our thinking in general to adopt disengaged standpoints on the world and our lives and to assess theoretical descriptions of reality based on supposed correspondence to reality rather than on their effectiveness in dealing with real-world dilemmas. Here there is considerable difference between the pragmatic and the deconstructionist approaches. The deconstructionist approach's central idea is that any theoretical description of reality will necessarily have some aporia since no such description can do justice to the relation to the other. By contrast, the pragmatic approach accepts that there can be legitimate theoretical descriptions of reality but adds that their litmus test is whether they satisfy conditions of justice, whether they embody our most inventive solutions to the crises we face in balancing our responsibilities.

What can I offer by way of support for this approach? The bulk of my argument in this book has appealed to textual evidence. This evidence may not be enough, however, to conclusively establish it; Levinas simply is not a systematic philosopher. There are some philosophers, such as Kant, who produce writings that are by no means clear but who are cautious thinkers, and debates over how to interpret them can be settled through careful exegesis of their writings. Levinas does not fall into this category. One gets the sense that he saw his writing as an opportunity to engage in exploration, to shed light on a dimension of our lives that he finds vastly important yet difficult to discuss, and he is willing to employ a variety of rhetorical strategies to illuminate it, even if doing so leads him to make pronouncements that are opaque or paradoxical. It may not be possible to settle interpretative debates, in turn, in the same way that we do with other philosophers, through careful analysis of selected passages, and a responsible exposition of his work should admit that such closure may not be available to us, that a given passage may support multiple readings. With Levinas, the task consists more in identifying insights that recur over the entirety of his corpus, and developing an overall interpretation that explains how they cohere with one another. I believe that there is textual evidence for my pragmatic approach, but a better argument would be if I could establish that it resonates with the overall tone of his philosophy better than competing interpretations.

Here, then, is my first broader argument. The deconstructionist reading is not morally robust enough to do justice to the role that ethics plays in Levinas. Now, some have criticized deconstructionism in general for being amoral. Its early proponents tended to champion it as a relativistic, even nihilistic approach to literary criticism, and this presentation led critics on both the right (Allan Bloom) and the Left (Terry Eagleton) to denounce it as morally impoverished. My complaint is different. Recent defenses of deconstructionism, many of which draw on Levinas in documenting an ethical turn in it, have done an admirable job of showing how deconstructionist critical strategies flow out of a certain ethical sensibility and are consistent with forms of ethical and political critique. So, my worry is not that it is inappropriate to link Levinas with deconstruction because deconstruction is amoral or nihilistic. My worry is that even the brand of deconstruction advanced by proponents of the ethical turn has too thin a conception of ethics to do justice to the significance Levinas attaches to it.

Dictums about the importance of ethics pervade Levinas's writings. He repeatedly pronounces that ethics is "first philosophy," and there is the frequent suggestion by him that responsibility for the other pervades all areas of our lives. The deconstructionist approach recognizes that he uses this terminology but points out that he has something unique in mind by "ethics." For Levinas, "ethics" does not mean a system of moral norms, rules of conduct, or the branch of philosophy that studies values such as "good" and "evil." It refers to the relation to the other, a relationship that, he tells us, overflows comprehension. Similarly, Levinas uses "responsibility" to refer not to fulfilling one's *obligations* but, as the term's etymology suggests, to being *open* to the other, to being prepared *to respond*, to existing, in Levinas's French, "in reply" (*repondre*) to the other. The proponent of the deconstructionist approach recognizes, then, that Levinas stresses ethics and responsibility but cautions against simplistic interpretations of these terms. For he uses them to refer to something quasi-epistemic—not so much acting on the basis of moral norms but accepting the limits of comprehension, being open to the other in her full alterity, with all the insecurity that such openness demands.

It is true that Levinas uses "ethics" and "responsibility" idiosyncratically. Nonetheless, the deconstructionist approach bleeds too much out of the connotations of these words. Levinas does not use them to straightforwardly refer to moral norms, but he does not have something wholly unrelated in mind. Consider how he describes the relation to the other in the following quotations:

> This gaze that supplicates and demands, that can supplicate only because it demands, deprived of everything because entitled to everything, and which one recognizes in giving (as one "puts the things in question in giving")—this gaze is precisely the epiphany of the face as a face. The nakedness of the face is destituteness. To recognize the Other is to recognize a hunger. To recognize the Other is to give.[28]

> The dimension of the divine opens forth from the human face. A relation with the Transcendent free from all captivation by the Transcendent is a social relation. It is here that the Transcendent, infinitely other, solicits us and appeals to us. The proximity of the Other, the proximity of the neighbor, is in being an ineluctable moment of the revelation of an absolute presence (that is, disengaged from every relation), which expresses itself. His very epiphany consists in soliciting us by his destitution in the face of the Stranger, the widow, and the orphan.[29]

The relation to the other is ethical in a robust sense: the face-to-face encounter seems to involve experiences and values that we customarily associate with ethics, such as pity for those who are vulnerable, accepting responsibility for their well-being, acts of charity, and so forth. No doubt Levinas has an idiosyncratic understanding of the terms "ethics" and "responsibility." He is explicit in warning his readers not to hastily equate these terms with systems of moral rules. Nonetheless, he also uses them to refer to values and ideas that are straightforwardly ethical. The relationship to the other seems to be not merely quasi-epistemic, a relation to the unknowable, but robustly ethical in that it involves pity, accepting responsibility for others, self-sacrifice, and generosity.

Nor is this description of the relation to the other emblematic only of Levinas's early thinking. Similarly robust passages can be found in *Otherwise than Being*:

> The response of the responsible one does not thematize the diachronical as though it were retained, remembered or historically reconstructed. It can not thematize or comprehend. Not out of weakness; to what could not be contained there corresponds no capacity. The non-present is in-comprehensible by reason of its immensity or its "superlative" humility or, for example, its goodness, which is the superlative itself.[30]

> This difference in proximity between the one and the other, between me and the neighbor, turns into non-indifference, precisely into my responsibility. Non-indifference, humanity, the one-for-the-other is the very signifyingness of signification, the intelligibility of the intelligible, and thus reason. The non-indifference of responsibility to the point of substitution for the neighbor is the source of all compassion. It is responsibility for the very outrage that the other, who qua other excludes me, inflicts on me.[31]

Levinas's aim in the second quotation is to clarify the relationship to the other, what he terms "proximity" in later writings, and his suggestion is that this relation is defined by an almost saintly capacity for caritas, unlimited charity and kindness, a total refusal of apathy toward the other, even to the point of accepting responsibility for the harms she does to us.

In the first quotation, he takes up the question of why this proximity cannot be comprehended, and his suggestion is that its incomprehensibility is not due to some failure to fully cognize it, a "weakness." Its incomprehensibility is due to the fact that it is not, properly speaking, an object of knowledge, a knowing, but "goodness," something that makes a claim on how we live.

Even in later writings, then, Levinas associates the relation to the other with values we typically associate with ethics. Whatever else it involves, the relation to the other has something to do with charity and kindness. The problem with the deconstructionist approach is that it undervalues these associations. A deconstructionist reader recognizes that the relation to the other involves responsibility. Yet she characterizes this responsibility in overly epistemic terms; starting with the idea that the other cannot be comprehended, she concludes that we have an obligation to reign in our efforts to totalize her and to remain vigilant against modes of thinking that violate her alterity. However, Levinas's characterization of this relationship seems to involve responsibilities that go beyond epistemic humility and critiquing totalizing modes of thinking, such as giving one's resources to help those who suffer and acting on their behalf.

This brings me to my second broader argument for the pragmatic interpretation. I began this book by considering the criticism that Levinas leaves our thinking fatally undetermined on matters of ethical significance. He talks about ethics, but his analysis of the relation to the other renders it impossible for him to offer anything like a normative ethics. I have sought to show that nothing in his philosophy bars him from developing a normative ethics, indeed, that hints of one can be teased from his writings, and that he has reasonable grounds for not offering further clarification of his ethics. However, it seems to me that the original criticism is legitimate: it *would be a problem* if his philosophy made it impossible to develop positive claims about how to express the relation to the other in a system of moral norms. Moreover, this criticism applies to the deconstructionist reading.

The deconstructionist approach reads Levinas, again, as seeking to show how thought and social life are structured by a relation that cannot be comprehended but only indirectly suggested by drawing attention to how it generates aporias in them. However, if one begins with this assumption, that the relation to the other is radically inexpressible, then the conclusion that Levinas can offer no positive account of justice is unavoidable. A benefit of the pragmatic approach is that it avoids this conclusion. To be clear, nothing in my approach bars me from making similar observations as the deconstructionist about the need to continually monitor our normative ethical theories—to continually scrutinize whether they in any way truncate respect for the other. However, the pragmatic approach maintains that it is also possible to discuss whether one theory does a better job than another of helping us realize our responsibilities to others. It is not only possible to "deconstruct" justice into ethics but to positively generate justice from ethics.

I will defend this claim in a moment. First, however, let me return to a criticism I noted earlier. One could argue that I am misrepresenting the deconstructionist reading in associating it with the idea that the other is *ineffable*. Its claim could be that the relation to the other is *overdeterminate*; its ethical content is so great that it generates multiple, conflicting norms. Consider my stock example of a face-to-face encounter: passing by a stranger in need of help. One might object that the problem is that I find myself overwhelmed with too many responsibilities in finding myself responsible for the other. I experience a responsibility to give to her. Yet I may also have a responsibility to give to a charity organization that will more effectively distribute my resources. Or perhaps it is not a duty to give that I owe but to agitate on behalf of broader political reforms that will aid her. Or perhaps, paradoxically, my responsibility consists in refraining from helping; perhaps aiding her renders her reliant upon me, leaving her trapped in circumstances in which she will continue to be victimized. One could argue that the problem is precisely that *all* these responsibilities flow out of this encounter, even those that contradict one another, with the effect that is impossible to single out any subset of uniquely expressive ethical norms. Instead, all we can do is deconstruct accounts of justice that would boast such a claim.

I dwell on this criticism so as to draw attention to the implications of my second broad defense of the pragmatic approach. My claim is that the deconstructionist approach is flawed because it leads to the conclusion that one cannot give a positive account of justice. The pragmatic approach has an advantage in that it has strategies for determining whether one set of norms does a better job than others of manifesting our responsibilities for the other. However, if I can establish this point, then I will have also answered the criticism that I am confusing ineffability with overdeterminacy. For I will have shown that *even if* several competing, contradictory moral norms are determined by an encounter with the other, it does not follow that, for Levinas, we cannot assess which ones represent, all other things being equal, the fullest, most ideal expressions of the ethical content of this relation.

So, how are we to select between competing moral norms? I find it useful to suggest a comparison between the pragmatic approach and the remarks that Thomas Kuhn makes on the considerations to be used in selecting between scientific paradigms in *The Structure of Scientific Revolutions*. Now, I want to reiterate my earlier warning about associating too disparate an array of thinkers with pragmatism. Kuhn's work has been associated with pragmatism, but he himself did not, to my knowledge, use this term to describe his thinking. Nonetheless, one thing Kuhn shares with the pragmatists, and Levinas, is his antiverificationism—his insistence that it is mistaken to think that we select scientific theories by straightforwardly confirming them against experience and, instead, that we accept them when they display a range of merits that, taken together, allow them to solve dilemmas more efficiently than the theories they replace. Reviewing Kuhn's exceptionally clear remarks on how we choose between scientific paradigms in the absence of verification will allow me to indicate how a pragmatic reading of Levinas can adjudicate between competing ethical norms that purport to be expressive of the relation to the other.

Kuhn famously argues in *The Structure of Scientific Revolutions* that previous philosophers of science, such as Karl Popper, had been wrong to think that scientists directly verify their theories based on evidence gained from experimentation. Kuhn pointed out that they instead accept more sweeping "paradigms" that help explain and predict a broad range of phenomena, and they then work at limited "puzzle solving" within these paradigms.[32] Should researchers encounter anomalous evidence in the course of puzzle solving, evidence that does not bear out a paradigm's predictions, their reaction is typically not to assume that a paradigm has been refuted but to engage in further experimentation that seeks to resolve the anomaly. Should enough anomalies accrue, however, it may inspire researchers to radically revise existing scientific belief and practice. This new paradigm will replace its predecessor, however, not because it has been conclusively *proven* to be true, whereas its predecessor has been proven to be false, but because it possesses a range of merits that make it *preferable* to the paradigm it replaces, such as explaining the preceding paradigm's anomalies, establishing new puzzles for researchers to investigate, and being more parsimonious in explaining a broader range of phenomena using fewer assumptions.

Several of Kuhn's readers concluded that this analysis of scientific progress led to a form of scientific relativism. For there is the idea in Kuhn that evidence underdetermines theory: a given piece of evidence can support a range of incommensurable scientific paradigms. Yet if we cannot verify the accuracy of one paradigm over another, and if the existing evidence supports multiple, incommensurable paradigms, then it seems to follow that there can be no neutral, reasonable grounds for selecting one over another. Kuhn himself, however, rejected this conclusion. It may be impossible to directly verify one theory or refute others, but it does not follow that one cannot be found superior insofar as it possess virtues that distinguish it from the theories it replaces. As Kuhn puts it, "techniques of persuasion"[33] can be used to defend the reasonability of one paradigm over another, such as the former's success in solving problems that posed a crisis for the latter, displaying greater quantitative precision, and "aesthetic" considerations such as being more parsimonious.

Kuhn's reply to the charge of being relativistic interests me because it provides a model for how to select between competing theoretical perspectives in cases where evidence underdetermines theory, and a pragmatic reader of Levinas can use a similar model to assess moral theories. Even if aspects of the relation to the other render it in some sense inexpressible—inexpressible to the point where we cannot directly derive moral norms from it—or even if competing norms seem to flow out of an encounter with the other, we can still ask whether one set of norms has virtues that render it superior to others. Similar "techniques of persuasion" can be used in ethics. For example, one can discuss whether various moral intuitions are treated as anomalies within a given ethical theory. Consider, along these lines, the objective purport of moral conviction—our perception that acts such as, say, massacres or rape campaigns are objectively wrong. An ethical theory that does not validate this feature of moral phenomenology is not thereby shown to be false, but it seems to tell against it if it depicts these convictions as confused or as mere statements of preference. Conversely, it should tell in favor of an ethical theory if it illuminates how our disgust and condemnation of such acts goes beyond being a mere matter of preference.

Just as one scientific theory can replace another, then, by offering explanations for data that confounded previous theories, so too one ethical theory can replace another by illuminating intuitions that previous ethical theories threatened to explain away. The fact that Ptolemy's astronomical models faced difficulty in explaining precession of the equinoxes did not show they were false, but it did count in favor of Copernicus's that they not only matched Ptolemy's in predicting the positions of the planets but also explained this phenomenon.[34] Similarly, it should count in favor of an ethical theory if it can help us make sense of moral intuitions that contradict other theories or that are left unexplained by them.

To be clear, my point is not that such arguments prove that one ethical theory is right whereas others are false. They simply constitute grounds for thinking that it is more *effective* than others, that it does a better job of illuminating our moral lives to us, that it issues, as I put it in chapter 3, more effective proposals for how to convey our moral intuitions in our actions. Indeed, my argument in chapter 4 would be an example of how this strategy might be put into practice. My claim against Kantian accounts of justice was not that they are false. My point was rather that they underestimate certain significant moral intuitions. The significance we attach to caregiving seems to be an important datum in our moral lives, as is our sense that persons with disabilities should not be treated as morally invisible with respect to issues of distributive justice. The fact that Kantian accounts of justice downplay these intuitions tells against them and in favor of a more robust account of distributive justice, for, again, what we want from an ethical theory is insight into how to enact these intuitions in our lives, and Kantianism reveals itself to be a faulty tool for enacting these elements of the ethical encounter with the other in asking us to suppress them.

Similar arguments can be made for Kuhn's other techniques of persuasion. For example, we can have a coherent discussion about whether one ethical theory has greater precision than others. Early in this book, I noted that some have read Levinas as endorsing a form of moral anarchy, as suggesting that rules get in the way of kindness and that we should engage in random acts of charity. I pointed out that this reading seems to face difficulties when confronted with tough ethical dilemmas, cases where there are several courses of action and where the correct choice is unclear. What we want in such situations, it seems, are action-guiding procedures of the type found in Kantianism and utilitarianism. This is not to say that these ethical theories are *truer* than a situational ethics that advises unpremeditated compassion, or even that they are necessarily preferable, but only that their greater precision counts as an asset that may render them preferable in some cases. If what we want from an ethical theory, in the face of a given dilemma, are concrete predictions about how to act, then the exactitude of Kant's End in Itself test may count as an asset that renders it more effective than other normative ethical positions, such as, arguably, some forms of situational ethics, care-based feminism, virtue ethics, and communitarianism.

Now, my point in the preceding paragraphs is not to argue that Kantianism is superior to situational ethics, or even that Kuhn's measures for assessing scientific paradigms can be straightforwardly applied to ethics. My aim is to show, in a preliminary way, how a pragmatic interpretation of Levinas might begin to formulate positive claims about justice.

This interpretation denies that Levinas is committed to radical claims about the other's inexpressibility. Instead, it reads him as stressing the need to reject spectatorial modes of thinking and to adhere to an agent-focused perspective. On this reading, there can be satisfactory theoretical descriptions of the world, social life, and the relation to the other. The issue is simply whether they satisfy conditions of justice, whether they embody our best solutions to the problems we face in balancing responsibilities. The key issue is not one of *accuracy* but *efficacy*: Does a theory render me open to the other? Does it inspire me to act on her behalf? Does it provide me with guidance that culminates in actions that do, in fact, benefit her? The pragmatic interpreter here shares the deconstructionist's skepticism about whether we can directly *deduce* a normative ethics by reflecting on the relation to the other. For this stratagem commits the spectatorial mistake—what Rorty might call the realist error, or what Lyotard might call the error of pursuing a "mimesis of an essence of justice"—of assuming that the central task in ethics consists in getting our moral theories to correspond with some moral state of affairs, when it rather consists in actually *acting* on the other's behalf. However, it does not follow that we cannot have a coherent, constructive discussion about the merits of one theoretical "tool" over others in aiding us in this task,[35] nor does it follow that we cannot develop a list of virtues similar the ones Kuhn associates with effective *scientific* theories but that distinguish effective *ethical* theories—virtues that enable effective ethical theories to succeed in guiding us to help one another. For example, I argued at the end of chapter 3 that an ethical theory that offers greater precision in its recommendations for how to live is, all other things being equal, more effective than one that supplies us with vague or *ad hoc* recommendations. Perhaps one could argue as well that an approach to conflicts in ethics that acknowledges that not all people share a single conception of "the good" is, all other things being equal, more likely to be effective in resolving conflicts than one that requires both parties in a conflict to share the same conception of the good. Again, the point is not that having these characteristics establishes that one theory is *truer* than another; having such virtues simply renders a theory a more effective tool for guiding our behavior and for negotiating conflicts.

Clearly, more needs to be done to specify this list of virtues that distinguish, all other things being equal, efficacious ethical theories. It may be the case, too, that any list will need to remain incomplete. Nonetheless, it is a mark in favor of the pragmatic interpretation that it is willing to initiate this discussion. The problem with the deconstructionist approach is that it portrays the challenge of developing a positive account of ethics, or what Levinas would call justice, as hopeless, as necessarily totalizing, either because the relation to the other is ineffable or, on more sophisticated variations, because it is overdeterminate. One cannot single out one set of moral norms that expresses one's responsibilities to the other; one can only expose the hubris in existing accounts of justice by documenting aporias in them. The problem is that this negative approach leaves our thinking fatally undetermined on dilemmas it would be negligent to avoid. Yet the same criticism cannot be leveled against the pragmatic interpretation. It may be difficult to assess whether one set of moral norms is more effective than another, or to draw up an uncontroversial list of "virtues" that distinguish effective moral theories, but the pragmatic interpretation has at least the benefit that it does not treat these challenges as being in principle unsolvable, and it can offer guidance on how to begin to assess competing sets of moral norms.

ENDNOTES

1. Bernasconi, "Rereading Levinas," in *The Question of the Other: Essays in Contemporary Continental Philosophy*, ed. Charles Scott and Arleen Dallery (Albany: SUNY Press, 1989), 23–34. *See esp.* p. 23.
2. See the discussion published as "Questions and Answers," 88–89.
3. See Levinas, "Enigma and Phenomenon," 65–77.
4. See Levinas, *Totality and Infinity*, 291.
5. Taylor, "The Validity of Transcendental Arguments," in *Philosophical Arguments* (Cambridge, MA: Harvard University Press, 1995), 20–33. *See esp.* p. 20.
6. Levinas, *Totality and Infinity*, 29.
7. Ibid., 24, italics added.
8. Ibid., 25.
9. De Boer, *The Rationality of Transcendence* (Stanford: Stanford University Press, 1997), 2.
10. Ibid., 30.
11. Levinas, *Totality and Infinity*, 25.
12. Ibid., 28.
13. Levinas, *Totality and Infinity*, 25, italics added. Bernasconi draws attention to this line in "Rereading *Totality and Infinity*," in *The Question of the Other: Essays in Contemporary Continental Philosophy*, ed. Charles Scott and Arleen Dallery (Albany: SUNY Press, 1989), 23–34. *See esp.* p. 24.
14. Bernasconi, "Rereading *Totality and Infinity*," 27.
15. Ibid., 23.
16. Bernasconi, "Rereading *Totality and Infinity*," 34.
17. Ibid., 26.
18. Mike Morgan develops a sophisticated version of something like the transcendental interpretation in *Discovering Levinas* (Cambridge: Cambridge University Press, 2007).
19. Allow me to acknowledge, however, a criticism I noted in chapter 1. I said in chapter 1 that Levinas's critics and defenders both accept the ineffability hypothesis. One could argue that I am mistaken: the worry is not that the relation to the other is *ineffable* but that it is *overdeterminate*. One could argue as well that this misrepresentation recurs in my claim that the deconstructionist approach "embraces the cluster of assumptions I associate with the ineffability hypothesis." The point is not

that this relation cannot be expressed in either of the major philosophic discourses, empiricism or transcendentalism, but that it sustains *both*, even though they are contradictory. This criticism is worth addressing. For now, I find it useful to continue to write as if the deconstructionist approach is wedded to the ineffability hypothesis. (Indeed, it seems to me that many proponents of it do not distinguish ineffability from overdetermination.) Later in this chapter, though, I will address the criticism that I am confusing ineffability with overdetermination.

20. Levinas's similarity to pragmatism is discussed in Chantal Mouffe's *Deconstruction and Pragmatism* (London: Routledge, 1996). The impression one gets, however, from pieces in this text is that it is inappropriate to liken him to the pragmatism. He is instead associated with deconstruction, and the pieces collected in *Deconstruction and Pragmatism* stress the rivalry between these two perspectives. As far as I know, others have not suggested the connections between Levinas and pragmatism that I explore in this chapter. However, Hillary Putnam and Stanley Cavell, both of whom are associated with pragmatism, have begun to show interest in his thinking.
21. Here I should note a possible dissimilarity between Levinas and recent pragmatists. There is a relativistic element to recent pragmatism that is foreign to Levinas, and, conversely, there is a universalizing element to Levinas that is foreign to pragmatism. Whereas pragmatists tend to eschew all talk of ahistoric, cross-cultural truths, Levinas often suggests that the revelation of infinite responsibility that occurs in the encounter with the other is universal. My suggestion is that Levinas resembles the pragmatists in his rejection of spectatorial standpoints, in favoring reformist approaches in ethics and politics, and in accepting broadly pragmatic criteria for assessing between competing normative ethical theories. He reaches these conclusions, however, on the basis of meta-ethical considerations involving the finality of moral obligation, which, in turn, lead him to part ways with some pragmatists on the issue of whether responsibility for the other is culturally contingent. I discuss this difference between Levinas and the pragmatists in greater detail later in this chapter. (See pp. 166–169.) Simon Critchley also raises something like this Levinasian critique against Rorty in "Deconstruction and Pragmatism: Is Derrida a Private Ironist or a Public Liberal?" in *Deconstruction and Pragmatism*.
22. Rorty, "Solidarity or Objectivity?," in *Post-Analytic Philosophy* (New York: Columbia University Press, 1985), 5.

23. This characterization of Levinas as a pragmatist may seem to contradict my argument in chapter 2. There, I said that Levinas is a moral realist. Now I seem to be saying that he thinks the content and scope of our ethical obligations can shift over time. There is no single, objectively true normative ethics or account of justice. One may render us maximally responsive to the needs of others today; an alternative may prove more efficient tomorrow.

    It is not clear, however, that there is a contradiction. My suggestion is that Levinas upholds a realist *metaethics* but would approve of using pragmatic considerations in constructing our *normative* ethical theories. The comparison I drew earlier between my reading of Levinas's and Lyotard's interpretation may be helpful here. Like Lyotard, I read Levinas as drawing attention to how the relation to the other is intrinsically normative—how it constitutes an unconditional "prescription that there be prescriptions." On my reading of the murder argument, Levinas attempts to show through it that it is impossible to consistently avoid or refuse this demand. It is in this sense that he is a realist: the prescriptive content of the relation to the other is not open to question, and it obligates me to insist on the reality of moral prescriptions in my life and in the world generally. Where I part ways with Lyotard is in his pessimism about the possibility of determining what form these prescriptions might take. Lyotard reads Levinas as drawing attention to how we cannot specify them. By contrast, I argue that Levinas would have us use pragmatic considerations to ensure that the moral commands we enforce in our lives do, in fact, serve the other.
24. Levinas, *Totality and Infinity*, 21.
25. Ibid., 22.
26. Ibid., 23.
27. This claim may be confusing. My point is that Levinas would not be troubled by the disinterested pursuit of a domain of knowledge if it could be shown that it helps us attend to our obligations to others. For example, there may be nothing wrong with the disinterested pursuit of knowledge in the sciences because "pure" research agendas uncover information that can be of great benefit to humanity.
28. Levinas, *Totality and Infinity*, 75.
29. Ibid., 78.
30. Levinas, *Otherwise than Being*, 11.
31. Ibid., 166.

32. I am being a little reckless in my brief overview of Kuhn. Kuhn's notion of scientific "paradigms" is notoriously slippery, and he himself said that it was the least well-understood area of his thinking. Briefly, Kuhn associated paradigms with what were accepted as exemplary instances of scientific research within the scientific community—an entire gamut canonical experiments, studies, concepts, and so forth, which are recognized by a scientific community as exemplary. Much more could be said about this idea. However, it is adequate for my purposes to gloss over the details of Kuhn's full account of scientific progress.
33. Kuhn, *Structure of Scientific Revolutions* (Chicago: University of Chicago Press, 1962), 152.
34. Kuhn discusses this example in *Structure of Scientific Revolutions*, 68.
35. It may help to give another example. Consider the shift in psychology from dualistic explanations of mental phenomena to naturalistic ones. At one point, psychology was viewed as distinct from biology and neuroscience in that it sought to make sense of the human *mind*, not simply the *brain's* mechanics. Increasingly, however, psychologists have incorporated physiological and neurological studies into their field, using them to explain mental disorders such as schizophrenia or bipolar disorder in terms of imbalances in the brain's chemistry. Here, then, we have a theoretical description of certain forms of human behavior. What might each of these approaches, the deconstructionist and the pragmatist, say about this attempt to thematize the other?

    The deconstructionist will seek to expose contradictions in naturalistic accounts of the mind and to show that they misjudge the complexity of the relation to the other. The proponent of the pragmatic approach can accept this point: she too finds it valuable to ask if naturalistic theories diminish respect for others. She may reach similar conclusions too about their reductiveness. Yet she can also ask if there are circumstances where they help us serve the other, and she can point out that they have proven valuable in helping to develop more effective treatments for schizophrenia. This is not to say that naturalism is *true*, nor that dualism is *false*. The pragmatic reader simply thinks we can discuss whether one model is more effective than another in helping us solve certain problems. In short, she thinks we should put aside questions about the *accuracy* of our theories and replace them with therapeutic questions: How effective are they in helping us help others? Are there circumstances where favoring a naturalistic account of the mind aids us in helping the other? Are there other circumstances in which it inhibits us?

There is a sense, then, in which the pragmatic reading of Levinas is more metaphysically pluralistic than the deconstructionist. For the deconstructionist, the goal is always to reestablish the same metaphysics: to show, once again, how our theoretical descriptions of the world and social life hover on a foundation they necessarily misrepresent. However, the pragmatic reader is willing to entertain any metaphysical stance, even to shift between contradictory perspectives, so long as they help her more effectively care for the other.

Conclusion

Levinas and Therapeutic Philosophy: Or, How to Throw Away a Ladder

Ludwig Wittenstein remarks at the end of his *Tractatus Logico-Philosophicus* that anyone who understands his arguments will "recognize them as nonsensical, when he has used them—as steps—to climb up beyond them." He adds that this reader "must, so to speak, throw away the ladder after he has climbed up it."[1]

There is debate over how to interpret this observation. One school of thought is that Wittgenstein is cluing his readers in to the therapeutic aim of his work. We tend to become tangled in confusions when we philosophize. We become, as he later puts it, "bewitched" by language. According to this school of thought, Wittgenstein's aim in *Tractatus* is not to advance some new metaphysics or new philosophy of language but to help us overcome our temptation to philosophize about the same old conundrums. Hence, his line about throwing away the ladder: he is cautioning his readers against taking his arguments too seriously. The whole point of them is to defuse his readers' temptation to philosophize, and Wittgenstein worries they will instead convert them into more fodder for this impulse. Hence his warning to remember that they are nonsense—philosophic nonsense intended to purge us of the temptation to talk more philosophic nonsense.

On this interpretation, Wittgenstein is part of a therapeutic tradition in philosophy that dates back at least as far as Kant, if not to ancient Greek skepticism. Kant famously begins his *Critique of Pure Reason* by noting that human reason has "this peculiar fate that in one species of its knowledge it is burdened by questions which, as prescribed by the very nature of reason itself, it is not able to ignore, but which, as transcending all its powers, it is also not able to answer."[2] Like Wittgenstein, Kant sees us as continually bewitching ourselves, and he too sets himself the therapeutic task of helping us work ourselves out of these confusions. There is also the suggestion in both Kant and Wittgenstein that these confusions result from a tendency on our part to confuse the theoretical and practical aspects of our lives. One of Wittgenstein's central ideas in his later writings is that we become entangled, when we philosophize, by questioning articles of belief which we could have no grounds to question in our actual, everyday lives. For his part, Kant argues that reason sets itself goals it can never satisfy through knowledge of the world, but only, he explains in his second *Critique*, through experiences we undergo in our ethical lives. For both, then, we have an all-but-perverse tendency to let ourselves become consumed with meaningless and unsolvable theoretical questions, and both find that the remedy is to draw their readers' attention back to the circumstances in which they make decisions in their practical, day-to-day lives.

One can summarize this book's argument by observing that I have tried to show that Levinas belongs in this therapeutic tradition. My intent in linking him with pragmatism is to establish his place in it. Wittgenstein, Kant, and contemporary pragmatists are all wary of our tendency to intellectualize the practical challenges we face in our lives, to replace real-world dilemmas with more palatable (if unsolvable) theoretical surrogates. Levinas operates in this tradition, although he differs from others in it that he stresses to a far greater degree the extent to which we seek to avoid our *ethical* responsibilities by replacing them with theoretical surrogates. His brand of philosophic therapy is issued on behalf of the other. His worry, unlike others in this tradition, is not so much that we harm *ourselves* by letting ourselves be bewitched into philosophizing, but that we exploit this mechanism in order to legitimate apathy and quietism. We let ourselves become bewitched into accepting detached, spectatorial perspectives because it lets us misrepresent our relationship to the other and the intimidating task of actually *acting* to help her.

Reading Levinas this way leads me to soften some of my earlier criticisms of the broadly deconstructionist approaches to Levinas which I have been concerned to criticize in this book. I began, in chapter 1, by considering Critchley's observation that Levinas does not offer guidance on how to express the relation to the other in a theory of justice. I argued in chapters 3 and 4 that he offers some guidance. However, it may be the case that this issue simply did not interest him. He seems to be far more troubled by our tendency to shirk our obligations by adopting spectatorial standpoints on the world and our lives. His primary concern is to root out this tendency. So, the deconstructionist is perhaps right to observe that Levinas does not offer much guidance on what account of justice he favors. She also may be right to fixate on worries about the totalizing nature of philosophy as the key concern of his work. It would be obtuse to deny that he spends considerable time in his writings identifying the many ways our customary modes of thinking falsify the relation to the other; so, the deconstructionist's focus on what I am calling the theoretical side of Levinas—his observations about the other's incomprehensibility—is understandable. Here I would only add that we differ in that I see Levinas as seeking to expose how we misrepresent *practical* challenges by treating them as if they were *theoretical*, whereas the deconstructionist treats him as making a cognitive-epistemic point about our failure to accept the limits on our ability to comprehend and to express the relation to the other.

I do worry, however, that the deconstructionist approach ends up perpetuating precisely that elision of the theoretical and practical that Levinas is concerned to critique. In Wittgenstein's terms, it refuses to throw away the ladder. If Levinas's primary aim is therapeutic, then I worry that the deconstructionist reading commits the same mistake that Wittgenstein worried his readers would make after finishing the *Tractatus*. Levinas's fear is that we too readily let ourselves become entangled in abstract, theoretical debates that have little bearing on the very real challenges we face in balancing our obligations. Instead of collaborating with you to find imaginative ways to resolve these challenges, I let myself be bewitched into yet another debate about whether I can know the other, whether I have objective moral responsibilities to her, whether I inadvertently "totalize" Levinas's thinking in seeking to explain it, and so forth. And yet, ever since Derrida's 1964 monograph, "Violence and Metaphysics," a majority of the material published on Levinas has focused precisely on such abstract issues, on issues such as whether he is consistent in his critique of philosophy as totalizing, whether he is consistent in his critiques of Husserl and Heidegger, or whether he violates his own claims about language. Yet if I am correct, the whole point of Levinas's critique of philosophy is precisely to move us to the point where we will see such questions as distracting, as, in Wittgensteinian terms, so much nonsense that prevents us from discussing how to help one another.

Put another way, the criticism that Levinas's philosophy is flawed because he does not say enough about justice applies more to *us*, his readers and commentators, than it does to him. For his part, Levinas could see himself as preparing the way for discussions of justice by rooting out the mechanisms we use to avoid them, the intellectual crutches we rely upon to falsify our relationship to the other. Their absence *in him* is understandable. Less understandable is their absence from so much of the scholarship that has emerged since *Totality and Infinity*.[3] If I am correct, then the key aim of Levinas's philosophy is to disabuse us of the temptation to become embroiled in epistemic debates over topics such as the limits of knowledge and whether our theories correspond to exteriority. His goal is to move us to the point where, instead, we will feel free to discuss justice—where we will accept that the main task of philosophic reflection should be to articulate creative proposals for how to best help others.

My suggestion, then, in arguing for a pragmatic reading, is that the time has come to throw away the ladder. The challenge we face consists precisely in avoiding the temptation to get embroiled in debates over totalizing philosophic theories, the other's inexpressibility, the irrecoverableness of "the saying" within "the said," whether Levinas's philosophy is shrewd or stupid for its supposedly unbridgeable "hiatus" between ethics and justice, and so forth. The challenge is to recognize these issues, in Wittgensteinian terms, as so much nonsense Levinas needed to discuss so as to move us, his readers, to the point where we can address dilemmas involving issues of justice that are far more urgent and far more complex—dilemmas such as feeding the hungry, clothing the naked, redressing grievances, sensitizing ourselves to subtle and unrecognized forms of humiliation, and reassessing what it is just to give to and to demand of others, what it is just to give to and to demand of ourselves, given the intricate and intimidating contexts in which we confront such issues in the world that we have inherited and in the conflicted space of our day-to-day lives.

ENDNOTES

1. See proposition 6.54 of Wittgenstein, *Tractatus Logico-Philosohpicus*, trans. D. F. Pears and B. F. McGuinness (London: Routledge, 1993), 74.
2. Kant, *Critique of Pure Reason*, trans. Norman Kemp Smith (New York: St. Martin's Press, 1965), 17.
3. Oddly enough, I here find myself in sympathy with claims Simon Critchley makes in his more recent work on Levinas. Critchley begins "Five Problems in Levinas's View of Politics" by lamenting the state of the scholarship on Levinas. He notes that "much of the work on Levinas tends to confine itself to exegesis, commentary, comparison with other thinkers, and, at its worst, homage." He adds that this state of affairs is "dull and produces only discipleship and scholasticism." See Critchley, "Five Problems in Levinas's View of Politics," 172–173. I share Critchley's frustration. My hunch is that he is right to link this state of affairs with the tendency among scholars of continental philosophy to confine themselves to exegesis. He is also correct to point out that Levinas scholars tend to lapse into "discipleship," into breathless admiration for Levinas. Finally, I think he is right call on Levinas scholars to refocus their attention on the difficult questions concerning ethics and politics that are raised by Levinas's writings. I am calling for the same shift in focus.

   I disagree with Critchley, however, on several points. First, his apparent suggestion is that the lamentable silence in Levinas scholarship on practical issues involving politics is partly due to Levinas's own silence on these issues. My view is that Levinas's silence has been exaggerated by critics. Moreover, his reticence is somewhat understandable for the reasons discussed above, whereas the silence in scholarship on him is, I argue, symptomatic of a misunderstanding of his philosophy. Second, my sense is that this silence is partly due to the ascendancy of deconstructionist readings of Levinas of the type that Critchley defends. On the one hand, such readings encourage us to focus on issues involving language, antifoundationalism, and the limits of comprehension and expression as the focal concern of Levinas's work. On the other hand, they discourage commentators from discussing ethical and political issues by depicting it as impossible to develop a positive account of justice. So, my sense is that Critchley is complicit in the trend which he criticizes, Having said this, "Five Problems in Levinas's View of Politics" is a step in the right direction. While I distrust Critchley's criticisms of Levinas, I admire his challenges to Levinas scholarship and his efforts to move the discussion in this scholarship on to more pressing political issues.

Bibliography

Affeldt, Steven. "The Ground of Mutuality: Criteria, Judgment and Intelligibility in Stephen Mulhall and Stanley Cavell." *European Journal of Philosophy* 6, no. 1 (1998): 1–31.

Alford, Fred. "Levinas and Political Theory." *Political Theory* 32, no. 2 (2004): 145–171.

Anselm. *Basic Writings*. Translated by S. N. Deane. La Salle, IL: Open Court, 1962.

Augustine. *Confessions*. Translated by R. S. Pine-Coffin. New York: Penguin Books, 1961.

Badiou, Alain. *Ethics: An Essay on the Understanding of Evil*. Translated by Peter Hallward. New York: Verso, 2001.

Bauman, Zygmunt. *Postmodern Ethics*. Oxford: Blackwell Publishers, 1993.

Becker, Lawrence. "The Finality of Moral Judgments." *Philosophical Review* 82 (1973): 364–371.

Benhabib, Seyla. *Situating the Self: Gender, Community, and Postmodernism in Contemporary Ethics*. London: Routledge, 1992.

Bernasconi, Robert. "What is the question to which 'substitution' is the answer?" In *The Cambridge Companion to Levinas*, edited by Simon Critchley and Robert Bernasconi, 234–251. Cambridge: Cambridge University Press, 2002.

———. "Rereading Totality and Infinity." In *The Question of the Other: Essays in Contemporary Continental Philosophy*, edited by Arleen B. Dallery and Charles E. Scott, 23–34. Albany: State University of New York Press, 1989.

Bernasconi, Robert, and Simon Critchley, eds. *Re-Reading Levinas*. Bloomington, IN: Indiana University Press, 1991.

Bernstein, Richard. "Evil and the Temptation of Theodicy." In *The Cambridge Companion to Levinas*, edited by Simon Critchley and Robert Bernasconi, 252–267. Cambridge: Cambridge University Press, 2002.

Berube, Michael. *Life as We Know It: A Father, a Family, and an Exceptional Child.* New York: Random House, 1996.

Berthoz, S., J. L. Armony, R. J. R. Blair, and R. J. Dolan. "An fMRI Study of Intentional and Unintentional (Embarrassing) Violations of Social Norms." *Brain* 125 (2002): 1696–16708.

Calvino, Italo. *Invisible Cities*. Translated by William Weaver. San Diego, CA: Harcourt Brace and Company, 1974.

Cavell, Stanley. *Conditions Handsome and Unhandsome: The Constitution of Emersonian Perfectionism*. Chicago: University of Chicago Press, 1990.

Caygill, Howard. *Levinas and the Political*. London: Routledge, 2002.

Chalier, Catherine. *What Ought I to Do?* Translated by Jane Marie Todd. Ithaca, NY: Cornell University Press, 2002.

Chanter, Tina. *Ethics of Eros: Irigaray's Re-reading of the Philosophers*. London: Routledge, 1995.

Cohen, Richard. *Elevations: The Height of the Good in Rosenzweig and Levinas*. Chicago: University of Chicago Press, 1994.

———. *Ethics, Exegesis, and Philosophy*. New York: Cambridge University Press, 2001.

———, ed. *Face to Face with Levinas*. Albany: State University of New York Press, 1986.

Critchley, Simon. *The Ethics of Deconstruction: Derrida and Levinas*. Edinburgh: Edinburgh University Press, 1999.

———. "Five Problems in Levinas's View of Politics and the Sketch of a Solution to Them." *Political Theory* 32, no. 2 (2004): 172–185.

———. *Infinitely Demanding: Ethics of Commitment, Politics of Resistance*. New York: Verso, 2007.

———. "Metaphysics in the Dark: A Response to Richard Rorty and Ernesto Laclau." *Political Theory* 26, no. 6 (1998): 803–817.

Critchley, Simon, and Robert Bernasconi, eds. *The Cambridge Companion to Levinas*. Cambridge: Cambridge University Press, 2002.

Darwall, Stephen. *Philosophical Ethics*. Boulder, CO: Westview Press, 1998.

De Boer, Theodore. *Rationality as Transcendence*. Amsterdam: J. C. Gieben, 1997.

Derrida, Jacques. *Adieu to Emmanuel Levinas*. Translated by Michael Naas. Palo Alto, CA: Stanford University Press, 1999.

———. "Violence and Metaphysics." In *Writing and Difference*. Translated by Alan Bass, 79–153. Chicago: Chicago University Press, 1978.

Dostoevsky, Fyodor. *The Brothers Karamazov*. Translated by Ralph Matlaw. New York: W. W. Norton and Company, 1976.

———. *Crime and Punishment*. Translated by Constance Garnett. New York: Vintage Books, 1950.

Douglass, Frederick. *Narrative of the Life of Frederick Douglass, An American Slave*. New York: Penguin Books, 1982.

Foucault, Michel. *The History of Sexuality, Volume I*. Translated by Robert Hurley. New York: Random House, Inc., 1978.

———. *Madness and Civilization*. Translated by Richard Howard. New York: Random House, Inc., 1965.

———. *The Order of Things*. New York: Random House, Inc., 1970.

Gilligan, Carol. *In a Different Voice*. Cambridge, MA: Harvard University Press, 1993.

Greene, J. D., R. B. Sommerville, L. E. Nystrom, J. M. Darley, and J. D. Cohen. "An fMRI Investigation of Emotional Engagement in Moral Judgment." *Science* 293 (2001): 2105–2108.

Grossman, Vasilii. *Life and Fate*. Translated by Robert Chandler. New York: Harper Collins, 1986.

Habermas, Jurgen. *The Philosophical Discourse of Modernity*. Translated by Frederick Lawrence. Cambridge, MA: The MIT Press, 1987.

———. "Reconciliation through the Public Use of Reason: Remarks on John Rawls's Political Liberalism." *Journal of Philosophy* 92 (1995): 109–131.

Harman, Gilbert. "Ethics and Observation," In *Ethical Theory*, edited by James Rachels, 103–109. Oxford: Oxford University Press, 1998.

Heidegger, Martin. *Basic Writings*. Translated by David Farrell Krell. New York: Harper-Collins Publishers, 1977.

———. *Being and Time*. Translated by John Macquarrie and Edward Robinson. New York: Harper-Collins Publishers, 1962.

Hendley, Stephen. "From Communicative Action to the Face of the Other" *Philosophy Today* 40, no. 4 (1996): 504–530.

———. *From Communicative Action to the Face of the Other*. Lexington, KY: Lexington Books, 2000.

Herzog, Annabel. "Is Liberalism 'All We Need'? Levinas's Politics of Surplus." *Political Theory* 30, no. 2 (2002): 204–227.

Honneth, Axel. "The Other of Justice." In *Cambridge Companion to Habermas*. Edited by Stephen K. White, 289–324. Cambridge: Cambridge University Press, 1995.

Jonsen, Albert, and Stephen Toulmin. *The Abuse of Casuistry: A History of Moral Reasoning*. Berkeley, CA: University of California Press, 1988.

Kant, Immanuel. *Critique of Practical Reason*. Translated by Lewis White Beck. New York: Macmillan Publishing Company, 1993.

———. *Critique of Pure Reason*. Translated by Norman Kemp Smith. New York: Saint Martins, 1965.

———. *Groundwork of the Metaphysics of Morals*. Translated by H. J. Patton. New York: Harper and Row, 1964.

———. *The Metaphysics of Morals*. Translated by Mary McGregor. Cambridge: Cambridge University Press, 1996.

Kiekegaard, Soren. *Concluding Unscientific Postscript*. Translated by L. M. Sweason and W. Lowrie. Princeton, NJ: Princeton University Press, 1941.

Kittay, Eva. "Human Dependency and Rawlsian Equality." In *Feminists Rethink the Self*, edited by Diana T. Meyer, 219–266. Boulder, CO: Westview Press, 1997.

———. *Love's Labor: Essays on Women, Equality, and Dependency*. London: Routledge, 1999.

Korsgaard, Christine. *The Sources of Normativity*. Cambridge: Cambridge University Press, 1996.

Kripke, Saul. *Wittgenstein on Rules and Private Language: An Elementary Exposition*. Cambridge, MA: Harvard University Press, 1982.

Levinas, Emmanuel. *Alterity and Transcendence*. Translated by Michael B. Smith. New York: Columbia University Press, 1999.

———. "As if Consenting to Horror." *Critical Inquiry* 15 (1989): 484–488.

———. *Basic Philosophic Writings*. Translated by Simon Critchley and Robert Bernasconi. Bloomington, IN: Indiana University Press, 1996.

———. *Collected Philosophic Papers*. Translated by Alphonso Lingis. Pittsburgh: Duquesne University Press, 1987.

———. "Dialogue with Emmanuel Levinas." In *Face to Face with Levinas*, edited by Richard Cohen, 11–23. Albany: State University of New York Press, 1986.

———. *Difficult Freedom: Essays on Judaism*. Translated by Sean Hand. Baltimore: Johns Hopkins University Press, 1990.

———. *Entre Nous: On Thinking-of-the-Other*. Translated by Michael B. Smith and Barbara Harshav. New York: Columbia University Press, 1988.

———. *Ethics and Infinity: Conversations with Philippe Nemo*. Translated by Richard Cohen. Pittsburgh: Duquesne University Press, 1985.

———. *Existence and Existents*. The Hague: Martinus Nijhoff, 1978.

———. *Is it Righteous to Be?* Edited by Jill Robbins. Stanford, CA: Stanford University Press, 2001.

———. *The Levinas Reader*. Translated by Sean Hand. Oxford: Basil Blackwell, 1989.

———. *Nine Talmudic Readings by Emmanuel Levinas*. Translated by Annette Aronowicz. Bloomington, IN: Indiana University Press, 1990.

———. *Of God Who Comes to Mind*. Translated by Bettina Bergo. Stanford, CA: Stanford University Press, 1998.

———. *On Escape*. Translated by Bettina Bergo. Stanford, CA: Stanford University Press, 2003.

———. "The Paradox of Morality," in *The Provocation of Levinas*. Edited by Robert Bernasconi and David Wood, 168–180. London: Routledge, 1991.

———. "The Primacy of Pure Practical Reason" In *Man and World: An International Philosophical Review*, translator unknown, 445–453. 1994.

———. *Proper Names*. Translated by Michael Smith. Stanford, CA: Stanford University Press, 1996.

———. *Time and the Other*. Translated by Richard Cohen. Pittsburgh: Duquesne University Press, 1987.

———. *Totality and Infinity*. Translated by Alphonso Lingis. Pittsburgh: Duquesne University Press, 1969.

Lyotard, Jean-François, and Jean-Loup Thebaud. *Just Gaming*. Translated by Wlad Godzich. Minneapolis: University of Minnesota Press, 1985.

Mackie, J. L. "The Subjectivity of Values." in *Ethical Theory*, 76–102. Oxford: Oxford University Press, 1998.

Margalit, Avishai. *The Decent Society*. Translated by Naomi Goldblum. Cambridge, MA: Harvard University Press, 1996.

Miller, James. *The Passion of Michel Foucault*. New York: Doubleday Publishing Group, 1993.

Moll, J., R. De Oliveirra-Souza, and P. J. Eslinger. "Morals and the Human Brain: A Working Model." *Neuroreport* 14 (2003): 299–305.

Moore, G. E. *Principia Ethica*. Cambridge: Cambridge University Press, 1965.

Morgan, Michael. *Beyond Auschwitz: Post-Holocaust Jewish Thought in America*. New York: Oxford University Press, 2001.

———. *Discovering Levinas*. Cambridge: Cambridge University Press, 2007.

———, ed. *A Holocaust Reader*. New York: Oxford University Press, 2000.

Mortley, Raoul. *French Philosophers in Conversation*. London: Routledge, 1991.

Mouffe, Chantal, ed. *Deconstruction and Pragmatism*. London: Routledge, 1996.

Nagel, Thomas. *The Possibility of Altruism*. Princeton, NJ: Princeton University Press, 1970.

New, Melvyn, Robert Bernasconi, and Richard A. Cohen, eds. *In Proximity: Emmanuel Levinas and the Eighteenth Century*. Lubbock, TX: Texas Tech University Press, 2001.

Noddings, Nel. *Caring: A Feminine Approach to Ethics and Moral Education*. Berkeley, CA: University of California Press, 2003.

Nussbaum, Martha. "The Future of Feminist Liberalism," *Setting the Moral Compass: Essays by Women Philosophers*. Edited by Cheshire Calhoun. New York: Oxford University Press, 2004.

Plant, Bob. *Wittgenstein and Levinas: Ethical and Religious Thought*. London: Routledge, 2005.

Platinga, Alvin. *Warranted Christian Belief*. New York: Oxford University Press, 2002.

Plato. *Sophist*. Translated by Nicholas P. White. Indianapolis: Hackett Publishing, 1993.

Prinz, Jesse. "The Emotional Basis of Moral Judgments." *Philosophical Explorations* 9, no. 1 (March 2006): 29–43.

Putnam, Hilary. "Levinas and Judaism." In *The Cambridge Companion to Levinas*, edited by Simon Critchley and Robert Bernasconi, 33–62. Cambridge: Cambridge University Press, 2002.

Railton, Peter. "Moral Realism." *Philosophical Review* 95, no. 2 (April 1985): 163–207.

Rawls, John. "Kantian Constructivism in Moral Theory: The Dewey Lectures 1980." *The Journal of Philosophy* LXXXVII (September 1980): 515–572.

———. *Political Liberalism*. New York: Columbia University Press, 1993.

———. *A Theory of Justice*. Cambridge, MA: Harvard University Press, 1971.

Robbins, Jill. *Altered Reading*. Chicago: University of Chicago Press, 1999.

Rorty, Richard. *Achieving Our Country: Leftist Thought in Twentieth-Century America*. Cambridge, MA: Harvard University Press, 1998.

Rowe, William. *Rationality, Religious Belief, and Moral Commitment*. Ithaca, NY: Cornell University Press, 1986.

Sanfey, A. G., J. K. Rilling, J. A. Aronson, L. E. Nystrom, and J. D. Cohen. "The Neural Basis of Economic Decision-Making in the Ultimatum Game." *Science* 300 (2003): 1755–1758.

Sartre, Jean-Paul. *Being and Nothingness*. Translated by Hazel Barnes. New York: Washington Square Press, 1956.

Schiller, Friedrich. *Schillers Werke: Nationalausgabe*. Vol. 1. Weimar: Hermann Böhlaus Nachfolger, 1940.

Schiller, Friedrich. *Über Anmut und Würde*. Stuttgart: Reclam, 1971.

Shklar, Judith. *Ordinary Vices*. Cambridge, MA: Harvard University Press, 1984.

———. *Political Thought and Political Thinkers*. Chicago: University of Chicago Press, 1998.

Singer, Peter. *Practical Ethics*. Cambridge: Cambridge University Press, 1979.

Stocker, Michael. "The Schizophrenia of Modern Ethical Theories." In *Virtue Ethics*, edited by Roger Crisp and Michael Slote, 66–78. New York: Oxford University Press, 1997.

Taylor, Charles. *Philosophical Arguments*. Cambridge, MA: Harvard University Press, 1995.

Wallace, R. Jay. *Normativity and the Will*. Cambridge: Cambridge University Press, 2006.

Wiesel, Eli. *Legends of Our Time*. New York: Schocken Books, 1968.

Williams, Bernard. *Ethics and the Limits of Philosophy*. Cambridge, MA: Harvard University Press, 1985.

Wittgenstein, Ludwig. *Tractatus Logico-Philosophicus*. Translated by D. F. Pears and B. F. McGuinness. London: Routledge, 1974.

Wyschogrod, Edith. *Emmanuel Levinas: The Problem of Ethical Metaphysics*. The Hague: Martinus Nijhoff, 1974.

Index

Made in the USA
Monee, IL
12 November 2026

0bee279e-24b5-4d9d-818c-6b34f5f84955R01